THE PARADOX EXPLORER

BY ROHAN AGGARWAL

While every precaution has been taken in the preparation of this book, the publisher assumes no responsibility for errors or omissions, or for damages resulting from the use of the information contained herein.

THE PARADOX EXPLORER

First edition. February 15, 2023.

Copyright © 2023 Rohan Aggarwal.

ISBN: 979-8215414064

Written by Rohan Aggarwal.

Contents

10) "From Lunar Beauty to Ashes: The Result of Nuking the Moon"

11) "The Wandering World: Earth's Journey Beyond the Stars"

12) "Stepping Beyond Our Solar System: The Miracle of Stellar Engines"

13) "From Creation to Destruction: The False Vacuum Theory"

14) "Shining a Light on Space Colonization: How We Can Build a Moon Base Now"

15) "The Final Frontier: The Moon's Journey to Earth and Beyond."

16) "The Enigma of Existence: Finding Your Purpose in Life"

17) "The Last Candle Before the Night of Black Dwarfs"

18) "Cosmic Origins: A Journey Through the Big Bang"

19) "Trapped in the Great Unknown: The End of Space Exploration"

20) "The Great Multitude: The Consequence of Overpopulation"

21) "Infection vs. Defence: The Ebola Virus and the Body's Response"

Acknowledgements

I would like to express my deepest gratitude to everyone who contributed to the creation of "The Paradox Explorer". Writing this book has been a journey of exploration and discovery, and I could not have done it without the support and encouragement of so many people.

First and foremost, I would like to thank my readers for their interest and curiosity in the subject of paradoxes. Your enthusiasm has been a constant source of inspiration and motivation for me.

I would also like to acknowledge the editors, designers, and publishing team who worked tirelessly to bring this book to life. Their expertise and dedication have been instrumental in shaping this book into its final form.

Last but not least, I want to express my appreciation for the countless scholars and thinkers whose ideas and insights have contributed to the content of this book. The field of paradoxes is a rich and diverse one, and it has been an honour to learn from and build upon the work of so many brilliant minds.

Once again, I thank everyone who has contributed to this project in some way. I hope that "The Paradox Explorer" will continue to inspire and challenge readers for years to come.

Foreword

When I first began exploring the world of paradoxes, I was struck by the richness and complexity of this fascinating subject. As I delved deeper, I discovered that paradoxes are not just mind-bending puzzles, but also powerful tools for exploring the limits of human knowledge and understanding.

In "The Paradox Explorer," I have attempted to share this wonder and excitement with you, the reader. Through 50 stories that explore a diverse range of paradoxes, I hope to take you on a journey of discovery and exploration.

Each paradox is presented as a unique tale, complete with illustrative examples and historical context. Through these stories, I aim to not only explain the paradoxes themselves but also to demonstrate their relevance to our lives and the world around us.

As an author, my goal has been to make the study of paradoxes accessible and engaging to all readers, regardless of their background or level of expertise. I believe that the ideas and insights contained in this book have the potential to inspire and challenge readers in profound ways.

Above all, I hope that "The Paradox Explorer" will encourage you to question your assumptions, challenge your thinking, and explore the rich and complex world of paradoxes. May this book serve as a guide and companion on your journey of discovery.

"The Paradox of Extra-terrestrial Life: The Fermi Dilemma"

The Fermi Paradox is a mystery that has puzzled scientists and astronomers for years. The paradox states that despite the vast size of the universe and the presence of trillions of habitable planets, there has been no concrete evidence of extra-terrestrial life. This raises questions about the possibility of life in the universe and why we haven't made contact with any aliens.

The universe is estimated to be about 90 billion light-years in diameter and contains at least 100 billion galaxies, each with 100 to 1,000 billion stars. This leads to the conclusion that there should be ample opportunity for life to develop and exist. However, this is not the case, and we have yet to find any concrete evidence of extra-terrestrial life.

The reason for this could be due to the expansion of the universe. Even if there are alien civilizations in other galaxies, it is unlikely that we will ever know about them because they are beyond our reach. This leads us to focus our attention on the Milky Way, our home galaxy. The Milky Way consists of up to 400 billion stars and an estimated 20 billion sun-like stars, with a fifth of them having an earth-sized planet in its habitable zone. Despite this, we have yet to find any evidence of life beyond Earth.

The Milky Way is about 13 billion years old, and the first habitable planets were born about one to two billion years after its formation. Earth is only 4 billion years old, which means that there have been trillions of chances for life to develop on other planets. If just one of these planets had developed into a space-traveling super-civilization, we would have noticed by now. But, where are all the aliens?

One explanation for the paradox is the existence of "filters". These filters represent barriers that are difficult for life to overcome. There are two possibilities: the first is that we have passed great filters and that the process of life developing is much more complicated than we think. The second possibility is that the great filters are ahead of us and that life on our level is destroyed once it reaches a certain point.

Another possibility is that we may be alone in the universe. Despite our efforts to find extra-terrestrial life, the universe appears to be empty and dead. No one is sending us messages or answering our calls. This thought can be scary, but it is also a reminder of the importance of preserving life on our planet.

In conclusion, the Fermi Paradox remains one of the biggest mysteries of our time. Despite our efforts to find evidence of extra-terrestrial life, we have yet to find any concrete proof. This leads us to question the possibility of life in the universe and raises the possibility that we may be alone. However, the search for extra-terrestrial life continues, and the discovery of even one extra-terrestrial civilization would change our understanding of the universe forever.

"The Deep-Sea Blast: A Nuclear Experiment in the Trench"

What if the world's most powerful nuclear bomb was detonated at the deepest point of the ocean? This scenario is both intriguing and terrifying, as it raises questions about the potential consequences of such an event.

The Mariana Trench is the world's deepest known point, reaching a depth of about 11 kilometres, located at the edge of two tectonic plates. This area is known for its pitch-black conditions and extreme pressure, making it one of the last places on earth for humans to explore.

For the purpose of this scenario, we'll use the RDS-220 hydrogen bomb, also known as Tsar Bomba, which has an explosive power equivalent to 50 megatons of TNT. Upon detonation, a chain reaction of nuclear fuel would occur, releasing an enormous amount of energy in a fraction of a second. This energy would produce a flaming bubble of water vapor, radioactive nuclei, and the remains of any living creatures in the area.

At the surface of the earth, the fireball bubble would grow rapidly, but at the Mariana Trench, the pressure is too great, causing the bubble to shrink and eventually disintegrate into smaller, hot, and radioactive bubbles drifting upward. As the radioactive fallout reaches the surface, it would likely cause a small wave and a bubbling plume of radioactive water in the Pacific, but no tsunami would affect coastal cities. The radioactive fallout would eventually be diluted into the ocean, although some radioactive water and salt would reach the atmosphere and rain down again.

Despite the concerns about earthquakes and volcanoes being triggered by the explosion, the truth is that such an event would likely have little effect on the earth's orbit and seismic activity. Earthquakes are already common at tectonic plate boundaries and even with over a thousand nuclear tests conducted in the past 70 years, the earth's orbit has remained unchanged.

In conclusion, while the idea of detonating a nuclear bomb at the Mariana Trench may sound frightening, the reality is that the consequences would be minimal. The planet is simply too big to be affected by the strongest forces humanity can unleash. In this scenario, the impact would likely be limited to a small wave and a radioactive plume in the ocean, with no significant effect on the earth's orbit or seismic activity.

"The Ultimate Cosmic Eraser: How Black Holes Threaten the Universe"

The universe is a vast and mysterious place, filled with incredible phenomena that challenge our understanding of reality. One of the most powerful and intriguing objects in the cosmos are black holes. These are areas of intense gravity, where matter is compressed into a tiny space and its pull is so strong that not even light can escape. The very existence of black holes raises questions about the nature of the universe and what lies beyond the limits of our understanding.

At the heart of the black hole is a paradox, one that threatens to delete the very fabric of reality itself: the information paradox. According to quantum mechanics, information is indestructible and cannot be lost. It may change form, but it cannot be destroyed. However, black holes appear to destroy information, taking different things and making them the same. This creates the information paradox and poses a serious challenge to our understanding of the universe.

To understand the paradox, it's important to grasp the concept of information. Information is not a tangible thing, but rather a property of the arrangement of particles. The atoms that make up the world around us don't care if they are part of a bird or a rock or a cup of coffee. It's the information, the arrangement of those atoms, that gives each object its unique properties. Without information, everything in the universe would be the same.

So, what happens to the information that enters a black hole? There are several possible solutions to the paradox. The first is that information is lost forever, requiring us to throw out all our current laws of physics and start from scratch. This is a frightening but exciting prospect, as it would open up new avenues of exploration and discovery.

Another possibility is that the information is hidden, stored in a new universe that we cannot observe or interact with. Or, it could be that a piece of the black hole remains after its life cycle, an information diamond filled with an infinite amount of data.

A third option is that information is safe after all, not lost or hidden. Perhaps black holes store information in some way that we haven't yet discovered. This idea raises new questions about where black holes keep their information and how they manage the cosmic housekeeping of the universe.

The information paradox is a fascinating and complex issue that challenges our understanding of the world around us. It's a mystery that scientists are still working to unravel, and one that has the potential to reveal new truths about the universe and our place in it. Whether we eventually find a solution to the paradox or not, one thing is certain: black holes will continue to captivate and inspire us for generations to come.

"The Blue Whale's Surprising Cancer Immunity: Peto's Paradox Demystified"

Cancer is a perplexing disease that continues to challenge scientists in their pursuit of a cure. One of the mysteries they've encountered is the biological paradox known as Peto's Paradox. According to this paradox, large animals seem to be immune to cancer, which goes against the conventional wisdom that the bigger a being is, the more cancer it should have. But why is this so?

At the heart of this paradox is the nature of cancer itself. Our cells are protein robots made up of millions of parts that are guided by chemical reactions. These complex reactions create and dismantle structures, sustain a metabolism, and make almost perfect copies of themselves. But with so many reactions happening in so many networks over so many years, it's not a matter of if something will go wrong, but when. Mistakes can add up, corrupting the grandiose machinery, but our cells have kill switches that make them commit suicide. If these kill switches fail, a cell can turn into a cancer cell.

Most of these cancer cells are quickly slain by the immune system, but over time, a cell can accumulate enough mistakes to slip by unnoticed and begin to make more of itself. All animals have to deal with this problem, but it's more prevalent in humans and larger animals, who have more cells and a longer lifespan, increasing the chance of something going wrong. And yet, even blue whales, with 3,000 times more cells than humans, don't seem to get cancer at all.

To explain Peto's Paradox, scientists believe there are two main ways: evolution and hyper tumours. The first explanation is that as animals became bigger, they had to invest in better cancer defences to survive. Large animals have a higher number of tumour suppressor genes, which prevent critical mutations from happening or order the cell to kill itself if it's beyond repair. These suppressor genes make the cells of large animals more resilient to cancer, but it probably comes with a cost that researchers are still trying to determine.

The second explanation is that large animals may have more hyper tumours than we realize, which may prevent cancer from becoming a problem. Hypertumours are the tumours of tumours and occur when the cooperation between cancer cells breaks down. Normally, cells work together to form structures like organs, tissue, or parts of the immune system, but cancer cells are selfish and only work for their own benefit. If they're successful, they form tumours, but this requires a lot of resources and energy.

The tumour cells trick the body into building new blood vessels directly to the tumour, feeding the thing killing it. However, cancer cells are inherently unstable and can continue to mutate, leading to the formation of a hypertumuor. The newly mutated cells can cut off the blood supply to the original cancer cells, causing them to starve and die. This process can repeat over and over, which may prevent cancer from becoming a problem for large organisms.

In conclusion, Peto's Paradox continues to confound scientists as they search for answers to this perplexing disease. While the explanations for why large animals seem to be immune to cancer may not be complete, they offer some insight into the complexities of this disease and the role that evolution and hyper tumours may play in preventing it. Understanding these factors is essential in our ongoing quest to defeat cancer once and for all.

"The Search for Humanity's Roots: A Journey Through the Ages"

The origin of humanity is a story that spans millions of years, starting from the split of the tribe of hominini from the apes to the emergence of Homo genus 2.8 million years ago. The first humans had cousins of comparable intelligence and ability, with at least six different human species coexisting with one another. However, only Homo sapiens survived, with a small percentage of Neanderthal and other human DNA mixing with our species.

The early humans used tools and controlled fire 2.8 million years ago, which made food more nutritious, increased the length of days and winters less harsh. The control of fire also allowed early humans to hunt, scare away predators and live in small hunter-gatherer societies. They had cultures of their own, communicated with each other in a proto-language and buried their dead.

Around 50,000 years ago, there was an explosion in innovation, and tools and weapons became more sophisticated, allowing for closer cooperation and a more advanced culture. This is what really sets humans apart from other creatures, as our ability to cooperate flexibly in large groups, expand knowledge quickly and preserve it over generations. The human brain had evolved, allowing for abstract thinking abilities and language, which was crucial in the preservation and expansion of knowledge.

For the next 40,000 years, human life remained more or less the same, with little innovation. Our ancestors were survival specialists, with a detailed mental map of their territory, fine-tuned senses to the environment, knowledge of plants and animals, and a rich social life within their tribe. The rise of agriculture around 12,000 years ago changed everything, as it allowed humans to hoard food on a large scale for the first time, leading to the rise of civilization.

Agriculture allowed individuals to specialize in various fields, from breeding more resistant crops to inventing things, leading to the creation of a reliable and predictable food source. This gave rise to the first civilizations, with the growth of cities, writing systems, and trade, leading to the creation of governments, religious systems, and the rise of empires. The agricultural age allowed humans to form more complex societies, leading to the creation of the world we live in today.

The journey from the split from the apes to the creation of modern civilizations is a story of evolution, innovation, and survival. It is a story of our species' resilience, determination and the drive to survive, even in the face of adversity. Our ancestors were superior in many ways, with their superior physical and mental abilities, but it was their ability to cooperate and innovate that allowed us to become the dominant species on the planet. The story of human origins is not just a story of our past, but of who we are and where we came from, and it serves as a reminder of our potential and the limitless possibilities of our future.

"The Matrix Unplugged: Decoding Reality"

The Simulation Argument is a philosophical thought experiment that raises the question of whether reality as we know it is actually real or just a simulation. The idea is based on the notion that our understanding of the universe is limited by our senses and brain's ability to process information. As technology advances, it becomes possible to simulate entire universes, leading to the question of whether our reality is just a simulation created by a more advanced civilization.

According to the argument, there are five assumptions that need to be met for the simulation hypothesis to be true. The first assumption is that it's possible to simulate consciousness. Although the true nature of consciousness is unknown, it's assumed that a human brain could be simulated through computer operations. The second assumption is that technological progress will continue and that civilizations will eventually have unlimited computer power. The third assumption is that advanced civilizations won't destroy themselves before reaching the point of being able to run simulations. The fourth assumption is that super advanced civilizations will have the desire to run simulations. And finally, the fifth assumption is that if there are many simulations, you are likely to be living in one.

The idea of the Simulation Argument has been modified by philosopher Nick Bostrom, and it raises several questions about the nature of reality and consciousness. The concept of simulating a brain is based on the idea that the brain is a complex system of interactions between neurons, which could be replicated by computer operations. However, simulating an entire human consciousness and all of human history would require a computer with an impossible amount of processing power, except maybe it isn't impossible.

Assumption two is that technological progress will not stop anytime soon, and that there may be civilizations with unlimited computer power in the future. The concept of the Matrioshka Brain, a theoretical megastructure made up of billions of parts orbiting a star, is one possibility for this kind of computer. Other advanced technologies like future quantum computers could also have the potential to simulate many civilizations at once.

The third assumption is that advanced civilizations won't destroy themselves before they have the ability to run simulations. The universe is full of potential Great Filters, such as nuclear war, asteroids, climate change, or black hole generators, which can prevent life from advancing. If life is inherently self-destructive, then there won't be any simulations.

The fourth assumption is that super advanced civilizations will want to run simulations for reasons that are unknown to us. It's difficult to determine the motivations of beings that are as advanced as gods, and it's possible that running simulations could seem like a foolish idea to them.

Finally, the fifth assumption is that if there are many simulations, you are likely to be living in one. This assumption is based on the idea that if there are simulated civilizations, it's more probable that you are living in one rather than in the original reality.

In conclusion, the Simulation Argument raises several questions about the nature of reality and consciousness. While the five assumptions may be true, there is no way to know for sure whether our reality is real or just a simulation. The concept of the simulation hypothesis may seem unsettling, but it's a fascinating thought experiment that challenges our understanding of the universe.

"The Abyss Awaits: Exploring the Isolation of the Deep Sea"

The Deep Sea, the most solitary place on Earth, is a wet and deadly desert inhabited by mysterious creatures living in total darkness. Despite the vastness of the Earth's oceans, only 2% of the planet's biomass is located there, and 90% of that is found in the first 200 meters near the surface. This area is where most of our attention has been focused so far, with its familiar coastal waters, where we swim, fish, pollute, and conduct science. However, when we delve deeper into the remote waters, we eventually reach the edge of the continental shelf and are faced with the continental slope, a long descent down to the deep sea. With every additional meter of water, light fades, and there are fewer plants and photosynthesis.

At depths beyond 200 meters, we enter the twilight zone, where the pressure rises to dangerous levels. Yet many fish and other creatures spend at least half of their lives down here, where they can rest and recover during the day and travel at night to shallower waters to feed. Over 90% of the species that live in this deep environment use bioluminescence to create light and communicate, hunt, and confuse predators. Down here, they also rely on marine snow, which consists of dead plant or animal parts, faeces, shells, sand, and dust, that constantly sinks from the surface to the bottom of the ocean.

In this environment, fascinating battles occur between unlikely enemies, such as sperm whales and giant squids. As we delve deeper into the midnight zone, a place of utter darkness beyond 1,000 meters, food becomes scarcer, and life has to adapt and become more energy-efficient to survive. For example, the vampire squid floats through the water without motion, using its long and slender arms to catch food and save energy, while carnivorous fish have multiple sets of long and deadly teeth to catch their prey on the first strike.

At abyssal depths, below 3,800 meters, life moves at a slow pace, preserving every last bit of energy. Everything down here moves slowly and elegantly, with the only exception being when they must escape danger. At 4,000 meters, the ocean floor is a seemingly endless abyss, where light and life are a rare sight. This is the deep sea, a wet and deadly desert, where the unknown and mysterious creatures that inhabit it still have much to reveal to us.

In conclusion, the deep sea is a place of stark contrasts, where life and light fade into darkness, and where the unknown and mysterious creatures that inhabit it still have much to reveal to us. From the familiar coastal waters near the surface to the deep and dark abyssal depths, the deep sea is a fascinating and uncharted world, waiting to be explored.

"The Starry Inferno: Navigating the Universe's Most Dangerous Realms"

The universe is vast and full of mysteries, but there's one thing that stands out as the most dangerous of them all: Strange matter. This bizarre substance is so extreme that it bends the rules of the universe and could infect and destroy everything it encounters. It's found in the cores of Neutron Stars, which are the densest things that are not black holes.

Neutron stars are formed after a massive star explodes in a supernova. The core of the star collapses under its own gravity, causing the electrons to be pushed into the protons and merging them into neutrons. This makes neutron stars like giant atomic nuclei, the size of a city, but with the mass of our Sun.

The environment inside the core of neutron stars is so extreme that the rules of nuclear physics change, leading to the creation of strange matter. This substance is made of quarks, which are the building blocks of protons and neutrons. Quarks come in many types, but only two of them make stable matter: the 'up' and 'down' quarks. The forces inside neutron stars are so extreme that they could cause the protons and neutrons to deconfine, dissolving into a bath of quarks. This would create a substance called quark matter, which is made purely from quarks.

However, if the pressure inside a quark star is great enough, some of the quarks could turn into strange quarks. Strange quarks have bizarre nuclear properties, and they are heavier and stronger than other quarks. If enough strange quarks are present, they could create strange matter. This substance might be the ideal state of matter, perfectly dense, perfectly stable, and indestructible. It's so stable that it could exist outside neutron stars, making it even more dangerous.

Strange matter could infect everything it encounters, converting it into strange matter as well. Protons and neutrons would dissolve and become part of the quark bath, releasing energy and creating even more strange matter. The only way to get rid of it would be to throw it into a black hole.

One of the ways strange matter could enter our solar system is through strangelets, which are droplets of strange matter. Strangelets could be as dense as the core of a neutron star, and they could drift through the galaxy for millions or billions of years until they collide with a star or planet. If a strangelet were to strike Earth, it would immediately start converting it into strange matter, growing as it does so. Eventually, all the atoms making up Earth would be converted, making the planet a hot clump of strange matter the size of an asteroid.

Similarly, if a strangelet were to strike the sun, it would collapse into a strange star, eating through it like fire through a dry forest. This would make the Sun less bright, leading to Earth freezing to death.

In conclusion, strange matter is one of the most dangerous substances in the universe, and it's found in the cores of neutron stars. It's made of quarks, which are the building blocks of protons and neutrons, and it could infect and destroy everything it encounters. Strangelets, which are droplets of strange matter, could drift through the galaxy and collide with a star or planet, leading to catastrophic consequences. It's a good thing that strange matter is confined to neutron stars, but it's still a fascinating and potentially dangerous substance that we need to understand better.

"Beyond the Limits: Exploring the Capabilities of Quantum Computers"

For centuries, humans have been upgrading their technology, from sharp sticks to nuclear weapons, and from their brains to computer machines. The development of computers has been exponential since the 1960s, with the computer parts becoming smaller and more powerful. However, we are now reaching the physical limits of this process as computer parts are getting smaller and smaller, approaching the size of an atom. To understand the reason behind this, we need to understand the basics of a computer.

A computer is made up of simple components, such as data representation, data processing, and control mechanisms. These components are made up of modules, which contain logic gates, which contain transistors. Transistors are the simplest form of data processors in computers, acting as switches that can either block or allow the flow of information. This information is represented by bits, which can either be 0 or 1. When several bits are combined, they represent more complex information. Transistors are combined to form logic gates, which perform simple tasks, such as an AND Gate, which sends an output of 1 if all its inputs are 1, and an output of 0 otherwise.

With computer parts becoming smaller and smaller, reaching the size of only a few atoms, quantum physics is making things complicated. As transistors are getting smaller, electrons may transfer from one side of the transistor to the other through a process called Quantum Tunnelling. This means that traditional computers are not making sense in the quantum realm. To overcome this barrier, scientists are developing quantum computers, which use the unusual properties of quantum physics to their advantage.

Quantum computers use qubits instead of bits, which can also be set to either 0 or 1. A qubit can be any two-level quantum system, such as a spin and magnetic field or a single photon. In the quantum world, a qubit can be in any proportion of both states at once, a property known as superposition. However, as soon as you measure the value of the qubit, it collapses into one of the definite states. This property makes quantum computers game-changers, as four qubits in superposition can be in all 16 combinations of 0 and 1 at the same time. This number grows exponentially with each extra qubit, making quantum computers vastly more efficient than traditional computers.

Quantum computers also have another property known as entanglement, which is a close connection between two qubits, making them react to a change in the other's state instantly, no matter how far apart they are. This means that by measuring just one entangled qubit, you can deduce the properties of its partner without having to look.

In quantum computers, quantum gates are used to manipulate probabilities and produce another superposition as its output. Quantum computers set up some qubits, apply quantum gates to entangle and manipulate probabilities, then finally measure the outcome, collapsing superpositions to an actual sequence of 0s and 1s. This means that a quantum computer can perform all possible calculations at the same time, and by cleverly exploiting superposition and entanglement, they can be exponentially more efficient than traditional computers.

Quantum computers will not replace traditional computers in all areas, but they are vastly superior in some specific areas, such as database searching. To find something in a database, a traditional computer may have to test every single entry, while quantum computers algorithms need only the square root of that time, which can be a huge difference for large databases. The most famous use of quantum computers is in IT security, where they are capable of breaking current encryption systems.

In conclusion, quantum computers are the future of technology, as traditional computers are reaching their physical limits.

Quantum Computers: The Future of Technology

"From Lunar Beauty to Ashes: The Result of Nuking the Moon"

The moon has always been a subject of fascination, not just for its beauty but also for its potential impact on Earth. What if we were to detonate a powerful nuclear weapon on the moon's surface? What would be the consequences?

To answer this question, let's imagine a hypothetical 100 megaton thermonuclear warhead, which is about twice as powerful as the most powerful bomb ever detonated, and place several curious astronauts around the moon as observers.

As soon as we push the button, several things happen in the first few milliseconds. High explosives send a shockwave to a radioactive metal core, compressing it to the point of criticality, starting a nuclear fission chain reaction. The 100-million-degree plasma created in this stage sets off the second stage, where atomic nuclei fuse like they do in the core of a star.

One of the biggest differences between explosions in space and on Earth is the absence of atmosphere. On the moon, the fireball shines and releases a wave of silent heat, rushing outwards in all directions, but without an atmosphere and oxygen-rich air, there's no burning. The topsoil of the moon is made from silicate rock and metal, chewed to dust by eons of meteorite impacts, mixed with tiny traces of water. When heated by the explosion, X-rays from the fireball vaporize a thin cloud of rock from the lunar surface, while the dust inside the fireball melts into glass.

Any astronauts within 50 km of the explosion would be fried by the deadly radiation, which is not weakened by the absence of an atmosphere. While all this happens, the explosion hammers against the moon, transferring a tenth of the explosion energy into seismic waves, causing an intense moonquake. The moon, being much smaller than the Earth, will feel an inescapable violent shaking, no matter where the astronauts are standing. Compared to an earthquake of 7 on the Richter scale, this shaking could seriously damage or level any infrastructure on the moon.

Where the bomb exploded, the ground splatters like water when a rock strikes a pond. As the explosion pushes against the surface, it may excavate as much as 100 million cubic meters of dust and rock, forming a crater 1 km across, while the bedrock is pulverized to rubble. Debris is shot into the sky in every direction, and much of it never returns to the moon, flying off faster than escape velocity.

A flurry of micrometeorites have been cast off to explore the solar system, many of which will rain down on Earth, though few will be larger than pebbles. Any satellite, astronaut, or space station in the way would have a really bad time. The micrometeorites are launched at different speeds and angles, allowing them to spread all over the surface of the moon, punching through any curious astronauts who stand in their way.

Finally, the explosion comes to an end. On Earth, the fireball rises and forms a mushroom cloud as it reaches up, but on the moon, there's no atmosphere, and therefore, no mushroom. The larger the plasma gets, the cooler it becomes, and the less energy it must make things happen. Within seconds of pulling the trigger, the bubble reddens and fades from view. It would be visible from Earth, like a brilliant flash of light.

In conclusion, a nuclear explosion on the moon would be an eerie and terrifying sight. The absence of atmosphere would allow the explosion to expand without impediment, causing fatal amounts of radiation, intense moonquakes, and micrometeorites raining down on Earth.

"The Wandering World: Earth's Journey Beyond the Stars"

The night sky might seem peaceful, but the truth is that stars are constantly moving through the galaxy at incredible speeds. Although they're unlikely to collide with Earth, they could still cause us a lot of problems. Gravity is the force that attracts every piece of matter to every other piece of matter in the universe, and it becomes weaker as the distance between objects increases. The massive Sun makes up 99.75% of the mass in the solar system, which shapes the behaviour and orbits of everything else.

Billions of years ago, the solar system was chaotic, but over time, a stable balance emerged. Today, most planets, asteroids, and comets have settled into safe orbits, with the Oort cloud at the edge of the system. However, if another star were to come too close to us, its gravity would disrupt the orbits of everything in the solar system. Some 70,000 years ago, a red dwarf, brown dwarf binary system passed through the Oort cloud, causing an asteroid onslaught that could still be headed towards us in the next 2 million years. A bigger problem looms on the horizon in the form of Gliese 710, a red dwarf headed towards the solar system that could disrupt the orbits of millions of objects in the Oort cloud in a million years.

The galaxy is an intense place, and stars get close to each other regularly, so it's possible that a star could pass directly through the inner solar system. Although the chance of a star colliding with the Sun is astronomically unlikely, the odds of a star passing by close enough to eject Earth from the solar system are estimated to be around 1/100,000 in the next five billion years. If this were to happen with an average red dwarf, Earth would be kicked out of the solar system.

As the star enters the solar system, it would appear as a small, orangish dot in the sky that grows bigger and brighter over the course of several months. The sun would slowly shrink in the sky, causing the temperature to drop and the final winter of humanity to begin. As the Earth passes Mars' orbit, the average surface temperature would drop to -50°C, and by the time it reaches Jupiter's orbit, the temperature would drop to -150°C, lower than the coldest ever recorded temperatures in Antarctica.

The polar ice caps would grow, forests would freeze, and animals would die in droves. As the global infrastructure breaks down, people would huddle together indoors, burning what they could for warmth as the temperature continues to drop and they run out of food. Everybody living on the surface would be living on borrowed time.

In conclusion, although stars moving through the galaxy are unlikely to hit Earth, their proximity could still cause problems for us. The biggest threat comes from a star passing directly through the inner solar system, which could eject Earth from the solar system and cause the final winter of humanity. So, while the night sky might seem peaceful, the reality is far from it, and we must always be prepared for the worst.

"Stepping Beyond Our Solar System: The Miracle of Stellar Engines"

The universe is always in motion and our Milky Way galaxy is no exception. The stars in our galaxy orbit around the galactic centre, including our own Sun. Although it maintains a distance of approximately 30,000 light years from the centre, completing an orbit every 230 million years, the dance of stars in the Milky Way can be chaotic and dangerous. As a result, our solar system could be at risk of encountering a star going supernova or a massive object passing by, potentially showering Earth with asteroids. To avoid such potential dangers, an advanced civilization with Dyson sphere-level technology could build a megastructure to steer a star through the galaxy - a stellar engine.

To move the solar system, only the Sun needs to be moved. All other objects in the solar system are held in place by gravity and would follow the Sun wherever it goes. There are several potential designs for a stellar engine, two of which, grounded in current understanding of physics, could potentially be built.

The first design is the Shkadov thruster, a giant mirror that works on the same principle as a rocket. The Sun releases photons as solar radiation, which carry momentum and can be reflected to create thrust. The Shkadov thruster would need to be positioned over the Sun's poles to prevent it from accidentally burning or freezing Earth with too much or too little sunlight. With a parabolic shape that sends most of the photons around the Sun in the same direction, the thruster could potentially move the solar system about a hundred light years over 230 million years at full throttle.

The second design is the Caplan thruster, which functions like a traditional rocket. It is a large space station platform powered by a Dyson sphere that gathers matter from the Sun to power nuclear fusion. The thruster shoots out a fast jet of particles at nearly 1 percent the speed of light and another jet to push the Sun along like a tugboat. To gather fuel, the thruster uses electromagnetic fields to funnel hydrogen and helium from the solar wind into the engine. The Dyson sphere refocuses sunlight to heat small regions on the Sun's surface, lifting billions of tons of mass that can be collected for fuel. The collected helium is burned in thermonuclear fusion reactors, creating a jet of radioactive oxygen expelled at a temperature of nearly a billion degrees, which serves as the primary source of propulsion. To balance the engine and prevent it from crashing into the Sun, a jet of accelerated hydrogen is shot back at the Sun. The Caplan thruster could move the Sun by 50 light years in as little as a million years.

In conclusion, the potential for a civilization with advanced technology to build a megastructure to steer a star through the galaxy is a fascinating concept. The Shkadov thruster and the Caplan thruster are just two examples of the potential designs for a stellar engine. While the complexity and difficulty of building such a structure is substantial, the reward of moving the solar system out of harm's way and ensuring the safety of our planet for millions of years to come is truly awe-inspiring.

"From Creation to Destruction: The False Vacuum Theory"

In a vast and mysterious universe, there is a hidden force that could spell its ultimate demise - vacuum decay. The concept revolves around the principles of energy levels and stability in physics. Everything in the universe has a certain level of energy, and it always wants to reach its ground state, in which it has the lowest possible energy and is completely stable.

This is true for every system, including the universe itself, which is said to be governed by quantum fields. These fields give the universe its properties and dictate how particles behave and interact. One of these fields, the Higgs Field, gives particles their mass, which determines almost everything in the universe's interactions.

However, it's possible that the Higgs Field is not stable but instead metastable, meaning it appears to be stable but is not. In this scenario, it's a false vacuum. Imagine a ball in a valley, the ball representing the Higgs Field. The valley might not be the true ground state for the Higgs Field, and there might be an even deeper valley that it wants to reach. This means the Higgs Field has a lot of potential energy waiting to be released.

A spark, such as quantum tunnelling, could trigger the Higgs Field's potential energy to be released, leading to vacuum decay. The Higgs Field would crash into the lower energy state, releasing a massive amount of potential energy and pushing space around it to collapse into a sphere of the new, stable Higgs Field or true vacuum. This sphere would grow at the speed of light in all directions, eliminating everything in its path and ultimately destroying the universe.

This vacuum decay would change all of physics as the standard model would be overthrown and replaced by new, unknown physics. The energy level of the Higgs Field would change, affecting how fundamental particles behave, how atoms hold together, and even how chemicals react, making life as we know it impossible. The outlook is indeed grim if vacuum decay were to happen, and the only thing we can do is to hope that our current understanding of particle physics is wrong.

While vacuum decay is a fascinating and terrifying thought, it's only speculation based on our current understanding, and no one can say for certain whether it's a real thing or just a scary idea. It's also possible that the spheres of death have already started expanding, but the universe is so vast that they might not reach us for billions of years, or maybe never at all, due to the expansion of the universe.

In conclusion, vacuum decay is a hauntingly beautiful reminder of the fragility of the universe, and the power of the laws of physics to determine its fate. Whether it's real or just speculation, it's a reminder that in this vast and infinite universe, anything is possible, and we should never take the existence of life and the universe for granted.

"Shining a Light on Space Colonization: How We Can Build a Moon Base Now"

Human beings have always been fascinated by the idea of exploring the galaxy and traveling to other worlds. Although it may seem like a far-off dream, the truth is, we have the technology to begin our journey right now by building a moon base. NASA and private organizations estimate that this project could cost anywhere from 20 to 40 billion dollars, which is not a big investment in the grand scheme of things. The benefits of building a moon base would be enormous, as it would provide a new sandbox for developing cutting-edge technologies, exploiting unlimited resources, and starting a new space race.

Building a moon base would require going through the three phases of colonization that have taken place throughout history. The first phase, exploration, began 60 years ago with the Apollo missions and has been further advanced by satellites and rovers that have mapped and studied the moon. The second phase, building the first moon base, is what we're ready to embark on today. The first moon base would be a small and lightweight structure consisting of inflatable habitats that could accommodate a crew of no more than 12 people. The astronauts will be tasked with conducting experiments and gathering data to enable humans to stay permanently on the moon.

The next step would be to make the moon base self-sufficient and economically productive, which would lead to the third phase of colonization. This would involve private contractors who are looking to get rich by extracting resources from the moon, such as precious metals and water, which could be used as rocket fuel. The lunar soil also has the necessary ingredients to make concrete, which would be useful for constructing new structures and expanding the colony. In the future, it may also be possible to mine Helium-3, a promising isotope that could be used for nuclear fusion reactors, providing cheap and clean energy to Earth.

Building a moon base is not without its challenges, as the moon is not a hospitable place for living things. The harsh environment and lack of atmosphere make it difficult for humans to survive, but we have the tools and technology to overcome these obstacles. The benefits of a moon base are too great to ignore, as it would provide us with an exciting new frontier to explore and the opportunity to lay the foundation for future missions to other planets. The time has come for us to take this bold step and start our journey to the stars.

"The Final Frontier: The Moon's Journey to Earth and Beyond."

The Moon and Earth share a unique relationship where the Moon orbits around Earth due to the Earth's gravitational pull. The question of what would happen if the Moon were to crash into Earth is a topic that has fascinated people for ages.

To answer this question, we first need to understand why the Moon isn't already on a collision course with Earth. It all boils down to the concept of speed and the laws of physics. The Moon stays in its orbit due to its sideways motion, which is called an orbit. It's much like when you throw a ball in the air, but the Moon is orbiting at a much faster speed of 3600 km/h, which keeps it suspended in space.

However, if the speed of the Moon were to change, its orbit would be altered, and it would eventually crash into Earth. But the force required to change the speed of an object as massive as the Moon is beyond comprehension. Igniting billions of rocket engines on its surface wouldn't even make a dent. So, we'll use our imagination and cast a magic spell that slows down the Moon so that it changes its orbit and spirals towards Earth.

Let's fast forward one year from the moment the magic spell is cast. During the first few days, nothing much changes, and the only noticeable effect is the rise in tides due to the Moon's gravity pulling back on Earth. However, as the Moon draws closer to Earth, the tides get higher every day. By the end of the second month, the tides have grown to 10 meters, and coastal cities are flooded, disrupting global infrastructure and communications. In the third month, communication and navigation satellites are disrupted, and their orbits become warped.

As we approach the fourth and fifth months, the tides rapidly grow to 30 meters and will reach 100 meters in height in a few short weeks. The oceans recede hundreds of kilometres during low tide, exposing the continental shelf, while walls of water flood agriculture, houses, and skyscrapers during high tide. The tides have reached their maximum, and the water in the oceans can no longer absorb the energy. The Earth experiences massive earthquakes and tsunamis, which cause widespread devastation and claim the lives of billions of people.

The aftermath of the Moon's collision with Earth would be catastrophic. The Earth's axis would be altered, causing massive climate changes and rendering much of the planet uninhabitable. The impact of the Moon would create a massive crater and throw debris into space, which would eventually form a ring system around Earth. The Moon would become part of Earth, increasing its size and mass, but the Earth's surface would forever be scarred by the impact.

In conclusion, the collision of the Moon with Earth is a hypothetical scenario that would have disastrous consequences. The laws of physics dictate that it's not possible for the Moon to simply crash into Earth, and it would take an immense amount of force to change its orbit. But let's not forget, this is all just an exercise in imagination, and in reality, the Moon will continue to orbit Earth for billions of years to come.

"The Enigma of Existence: Finding Your Purpose in Life"

The question "What are you?" is a philosophical inquiry into the nature of human existence. At the core of this question is the idea that the human self is not just a physical body, but a complex, dynamic system of trillions of cells, each of which is made up of thousands of different proteins. But what is it that makes these cells into a person, and what happens to the person if cells are removed, added, or replaced?

One of the central mysteries of the human body is how it can remain so stable and consistent over time, despite the fact that most of its cells are constantly dying and being replaced. Over the course of a single seven-year period, most of the cells in the body will be replaced at least once, and by the time a person reaches old age, they will have cycled through a billion cells. At the same time, cells are also undergoing mutations and changes due to environmental influences, which can alter the genetic code and therefore change the person themselves.

Another aspect of the human body that challenges the idea of the self as a static entity is the fact that some cells can break away from the normal biological social contract and become immortal. This is the case with cancer cells, which put their own survival over the survival of the body they belong to. This raises the question of whether these cells are part of the person, or if they have become a separate entity.

Yet another complicating factor is the fact that many of the components of the human body, such as the viruses and bacteria that make up 8% of our genomes, were not originally human but have merged with us over time. This raises questions about what constitutes the human self, and whether the self is a static entity or a dynamic, ever-changing pattern.

The problem of who we are is not just a question about the body, but also a question about the mind. The human brain, like the rest of the body, is also constantly changing, with neurons undergoing mutations and changes over time. These mutations and changes mean that the mind is also in flux, and that the self is not a single, unchanging entity but a constantly evolving system.

Ultimately, the question of who we are is one of the greatest mysteries of human existence, and one that may never be fully answered. Our human brains evolved to deal with absolutes, but the fuzzy borders that make up reality are hard to grasp. Ideas like life and death, you and me, may not be absolute, but rather part of a fluent pattern that is lost in the strange and beautiful universe. The nature of the human self is a complex, dynamic, and ever-evolving question that may continue to challenge us for generations to come.

"The Last Candle Before the Night of Black Dwarfs"

The universe is a vast and mysterious place, full of secrets yet to be uncovered. However, one thing is certain, the universe will eventually come to an end. But before that happens, there are places that will persist for billions of years - white dwarfs. These celestial bodies are the remnants of dead stars and could potentially be humanity's last refuge before the universe perishes.

Stars have varying lifetimes, depending on their size. Massive stars burn bright and fast, exploding in supernovae just a few million years after their birth. But for 97% of all stars, their final fate is to become a white dwarf. Small stars, known as red dwarfs, slowly burn out over trillions of years until they transform into white dwarfs. Meanwhile, medium-sized stars like our sun go through a more fascinating process.

The sun fuses hydrogen into helium in its core, releasing an immense amount of energy through its gravity. This energy pushes outwards, maintaining the delicate balance of the star. But as the hydrogen in the core gets depleted, the sun will begin to burn helium into heavier elements. As this process takes place, the outer layers of the star will be shed, leaving behind its former core as a white dwarf.

White dwarfs are incredibly dense, with a surface gravity over 100,000 times greater than Earth's. They are about the size of Earth, but with half of their former mass, meaning a teaspoon of white dwarf matter would weigh as much as a car. Despite being extremely hot, with temperatures reaching up to 40 times hotter than the sun, white dwarfs are not very active. All the heat within them is trapped, and the only way it can escape is through radiation, which is so inefficient that white dwarfs will take trillions of years to cool down. This makes them a potential source of light and energy in a dying universe, shining as long as 100 billion years, ten billion times longer than the universe has existed so far.

However, even white dwarfs will eventually die. They will transform into black dwarfs, inactive spheres with no energy left to give, cold and dark to the point of being practically invisible. The universe will enter its final stage, heat death, leaving behind a cold and dark graveyard filled with black holes and black dwarfs scattered across trillions of light-years.

The ultimate fate of black dwarfs is uncertain. If the proton, a fundamental part of atoms, has a limited lifespan, black dwarfs will slowly evaporate over trillions of years. If the proton does not decay, black dwarfs will turn into spheres of pure ions via quantum tunnelling, traveling alone through the dark universe for all eternity.

While the end of the universe may seem bleak, it is so far away that it may as well not happen at all. Right now, we have the opportunity to exist in a universe filled with stars, light, and planets, with enough time to explore and marvel at their beauty. It's a privilege to be alive at a time when the universe is still full of wonder and potential, and we should cherish it while it lasts.

"Cosmic Origins: A Journey Through the Big Bang"

The Big Bang - the beginning of everything. Before the mid-20th century, the notion of an infinite and ageless universe was widely accepted among scientists. However, with the advent of Einstein's theory of relativity and Edwin Hubble's discovery of galaxies moving away from each other, the concept of the Big Bang emerged as the prevailing scientific theory. The accidental discovery of cosmic background radiation in 1964, along with other observational evidence, solidified the theory even further. With the help of advanced technology such as the Hubble telescope, scientists now have a better understanding of the Big Bang and the structure of the cosmos.

Contrary to popular belief, the Big Bang was not an explosion in the traditional sense. Instead, it involved the rapid expansion of space everywhere all at once, starting from an incredibly small size and growing to the size of a football within the blink of an eye. The universe doesn't expand into anything because it has no borders and there is, by definition, no "outside" of the universe. It's all there is.

At the time of the Big Bang, energy manifested as particles that existed only for the briefest of moments. Glowing with heat and density, these particles formed pairs of quarks that eventually became hadrons, such as protons and neutrons. However, most of these hadrons were unstable, leaving behind only the most stable particles. After one second had passed, the universe had expanded to a billion kilometres in diameter and was cool enough for most neutrons to decay into protons and form the first atoms, hydrogen.

Over the next few minutes, the universe cooled down and settled, allowing atoms to form out of hadrons and electrons, creating a stable and electrically neutral environment. This period was known as the Dark Age as there were no stars and the hydrogen gas prevented visible light from passing through. As the hydrogen gas clumped together under the force of gravity, stars and galaxies began to form, their radiation dissolving the stable hydrogen gas into a plasma that still permeates the universe today.

The exact details of what happened right at the beginning of the Big Bang remain a mystery. Our current tools break down and natural laws no longer make sense at this point. To fully understand the beginning of the universe, a theory that unifies Einstein's theory of relativity and quantum mechanics is needed - a task that many scientists are currently working on.

While many questions about the universe remain unanswered, what we do know is that it all started with the Big Bang and gave birth to particles, galaxies, stars, Earth, and ultimately, us. We are not separate from the universe but are instead a part of it, made up of dead stars and serving as the universe's way of experiencing itself. Let's continue to experience and explore the universe, until there are no more questions left to ask.

"Trapped in the Great Unknown: The End of Space Exploration"

Space travel has always been an exciting and adventurous journey for humanity, but it might end up becoming a prison for us if we're not careful. With every rocket launched and satellite deployed, we're creating a trap that's getting deadlier every year. The more we venture into space, the more we're at risk of limiting ourselves to the Earth. The reason behind this is the creation of space debris.

Getting something into space requires a lot of energy and speed. Satellites and space stations need to be placed in low Earth orbit, just a few hundred kilometres above the Earth's surface. This height is ideal for these objects to stay in orbit for centuries, but it's also the source of the deadly trap. Rockets are made of metal cylinders and contain large amounts of fuel. Once the fuel is spent, the empty tanks are dropped, and most of the discarded parts remain in orbit and add to the space junk.

Low Earth orbit is now a junkyard of spent boosters, broken satellites, and millions of pieces of shrapnel from missile tests and explosions. The debris is moving at a speed of up to 30,000 km/h, creating a dangerous web of deadly pieces of destruction. If a satellite is hit by even a pea-sized piece of debris, it would be destroyed instantly, releasing energy to vaporize the debris. A trillion-dollar global infrastructure network is also located in this danger zone, performing critical duties that are essential to the modern world.

The situation in orbit is getting worse as the number of satellites and the amount of junk in orbit is expected to grow tenfold in the next decade. The biggest concern is the possibility of a chain reaction that turns a lot of non-junk things into junk. If two satellites collide, they turn into clouds of thousands of small pieces that are fast enough to destroy more satellites. This could trigger a collision cascade, where each collision creates more bullets, accelerating the destruction of everything in orbit.

The worst-case scenario would be a debris field made of hundreds of millions of pieces, many too small to track, moving at 30,000 km/h. It would create a deadly barrier around Earth that might be too dangerous to cross. The loss of our space infrastructure would set back the technology we rely on daily, and our dreams of moon bases, Mars colonies, or space travel might be set back for centuries.

While it might not be too late to clean up our mess, the space industry is still growing fast, and occasional weapon tests don't help. There have been suggestions on how to remove as much deadly space junk as possible, but it's not an easy task. The clean-up process would require international cooperation and a lot of resources, but it's imperative if we want to continue exploring space.

In conclusion, space travel is a risky business, and we need to be careful not to create a prison for ourselves. The creation of space debris is a serious concern, and we need to act quickly to prevent a disaster that might set back our space exploration for centuries. The clean-up process might be challenging, but it's necessary if we want to continue venturing into space and exploring the unknown.

"The Great Multitude: The Consequence of Overpopulation"

Overpopulation is a phenomenon that has been the cause of much concern for many years now. The number of people on the planet has skyrocketed in the last few centuries, starting with 1 billion in 1800, to 2.3 billion in 1940, 3.7 billion in 1970 and reaching 7.4 billion in 2016. The growth in population has led to numerous predictions about the future, including mass-migration, overcrowded slums and megacities, diseases and pollution, chaos and violence over energy, water, and food, and a human species focused only on sustaining itself. The fear is that overpopulation will destroy our way of life, but is this just an ungrounded panic?

In the 1960s, the population growth rate reached an unprecedented level, leading to apocalyptic predictions. It was believed that the poor would keep procreating endlessly and overrun the developed world. The legend of overpopulation was born. But it turns out that high birth rates and population explosions are not permanent features of some cultures or countries, but rather a part of a four-step process the entire world is going through, the demographic transition. Most developed countries have already gone through this transition, while other countries are in the process of doing so.

Going back to the 18th century, the entire world, including Europe, was in the first stage of the demographic transition. At that time, Europe was worse off than many developing regions, suffering from poor sanitation, poor diets, and poor medicine. A lot of people were born, but many of them died just as fast, so the population hardly grew. Women had between 4 and 6 children, but only 2 of them would reach adulthood. The industrial revolution in the UK changed the living conditions of people for the better, going from being peasants to workers. The manufacture of goods was streamlined and became widely available, sciences flourished, transportation, communication, and medicine advanced, and the role of women in society shifted. This economic progress not only formed a middle class, but it also raised the standards of living and health care for the poor working population, leading to the second stage of the transition.

Better food supplies, hygiene and medicine, meant that people stopped dying as frequently, especially at a very young age. This resulted in a population explosion, doubling the UK's population between 1750 and 1850. In the past, families used to have many children because only a few of them were likely to survive, but that changed with the better living conditions. The third stage of transition was set in motion, where fewer babies were conceived, and population growth slowed down. Eventually, a balance was reached between the death rate and birth rate, resulting in a stable population. Britain had reached the fourth stage of the demographic transition, and many other countries went through the same process.

So why is the population still growing despite low birth rates? This is because the children born in the population explosion of the 70s and 80s are having kids themselves now. This has led to a noticeable spike in overall population, but they are having far fewer children on average than their parents. The average number of children born today is 2.5, compared to 5, 40 years ago. As this generation gets older and fertility declines further, the rate of population growth will continue to slow down. This is true for every country.

In conclusion, overpopulation is a complex and multifaceted issue that has far-reaching consequences for the planet and its inhabitants. It poses significant challenges to the sustainability of resources, the stability of ecosystems, and the well-being of people. Governments, organizations, and individuals must work together to find solutions to this pressing problem. This may include reducing birth rates through family planning and reproductive health education, improving access to education and economic opportunities, and addressing the root causes of poverty and inequality. Additionally, the world must transition to more sustainable patterns of consumption and production and invest in renewable energy and sustainable agriculture. With proactive and collective action, we can work towards a more equitable and sustainable future for all.

"Infection vs. Defence: The Ebola Virus and the Body's Response"

The Ebola virus is a tiny entity that has caused fear and panic across the world. It's a virus that has no ability to do anything on its own and can only survive and multiply by infecting cells. The human body has an immune system that is designed to prevent infections such as Ebola, but when this virus strikes, it can wreak havoc on the body's defence system.

Ebola attacks the immune system in several ways. It starts by targeting the dendritic cells, which are the brains of the immune system. The virus enters the dendritic cell and takes over its resources to build more Ebola viruses. This leads to the release of millions of viruses into the tissue and the disruption of the immune system, making it unable to react. The virus also infects the guard cells of the body, such as macrophages and monocytes, and tricks them into signalling to the blood vessels to release fluid into the body, causing internal bleeding.

At the same time, the virus is spreading and killing liver cells, leading to organ failure and more internal bleeding. All the mechanisms of the immune system that have evolved to handle infections work against the body, and the virus continues to spread. In a last-ditch effort, the immune system launches a cytokine storm, which is an S.O.S signal that causes it to launch all its weapons all at once. Although this hurts the virus, it also causes collateral damage, especially in the blood vessels, leading to serious dehydration, organ failure, and death. Currently, six out of ten infected people die from Ebola.

Ebola is a nasty virus, but it's important to keep it in perspective. The only way to get infected by Ebola is by coming into contact with the body fluids of a person who shows symptoms or from an infected bat. So, it's important to take precautions, but there's no need to panic. Ebola has killed 5,000 people since 2014, while the common flu kills up to 500,000 people each year, and malaria causes up to one million deaths each year. The most infectious thing about Ebola is the media hype around it, so it's important not to let yourself be scared.

In conclusion, the Ebola virus is a deadly entity that attacks the body's immune system, leading to serious consequences such as organ failure and death. However, it's important to keep things in perspective and not let fear control us. By taking precautions and avoiding contact with infected body fluids, we can stay safe and healthy. So, stay informed, stay vigilant, and stay calm.

"The Carnivore Conundrum: Is Meat Good or Bad for You?"

The debate on whether meat is healthy or not has been raging on for years. On one hand, meat has been a staple in the human diet for millions of years and has provided us with the necessary nutrients to survive. On the other hand, recent studies have linked meat consumption to various health risks such as heart disease, certain cancers, and an early death.

Biologically, humans need to eat for three reasons: for energy, to acquire materials to make our cells, and to get special molecules that our bodies cannot make on their own. Meat contains all the essential amino acids that our bodies need, as well as important minerals like iron, zinc, and essential vitamins, some of which are difficult to find in plants.

Fish is considered one of the healthiest meats to consume, thanks to its high content of polyunsaturated fatty acids like omega 3. These fatty acids can lower the risk of cardiovascular disease and support anti-inflammatory immune functions. Chicken is also a good option, as it is regarded as the meat with the fewest health risks.

Red meats, such as beef, veal, pork, lamb, horse, and goat, can be problematic if consumed in large quantities. Studies have shown that eating 100 grams of red meat every day increases the risk of diabetes by 19%, strokes by 11%, and colorectal cancer by 17%. However, these studies are case-control studies and are difficult to eliminate other factors.

Processed meats, such as bacon, ham, salami, sausages, and hot dogs, are the most dangerous for our health. They contain harmful chemicals, such as nitrates and nitrites, which can damage our DNA and lead to cancer. The World Health Organization reviewed 800 studies over 20 years and concluded that processed meat is strongly linked to an increased risk of colorectal cancer.

The life that our meat lived when it was still part of a living being also plays a role in our health. The use of antibiotics to prevent diseases in livestock can lead to the spread of antibiotic resistance. A diet high in meat from animals treated with antibiotics can increase the risk of developing antibiotic resistance.

In conclusion, meat itself is not dangerous for us. However, the health effects of meat consumption can vary greatly, depending on how it is prepared, what animal it comes from, and how much of it is consumed. If you are concerned about your health, it is best to consume meat in moderation, opt for fish or chicken, and avoid processed meats. To ensure optimal health, it is also important to maintain a balanced diet that includes plenty of fruits and vegetables.

"The Paradox of War and Peace: Understanding the Shift in Human Conflict"

War and violence have been a part of human history since the dawn of time. However, despite the current ongoing conflicts such as the violence committed by ISIS, the invasion of Ukraine by Russia, and the ongoing conflict between Israelis and Palestinians, there is a paradoxical trend that suggests that war is slowly becoming a thing of the past. While the global population is at an all-time high, the numbers suggest that war is going out of fashion and that we are living in the most peaceful period in human history.

To understand this paradox, it's important to look at the current state of conflict. As of September 2014, there were only 4 conflicts happening in the world that had caused at least 10,000 deaths since January 2013. Although these conflicts are still a cause for concern, it's important to note that none of these conflicts are active wars between countries. Instead, they are either civil wars or local conflicts. Civil wars, while terrible and causing immense suffering, have a much smaller impact compared to wars between nations or empires. Wars between nations have the potential to mobilize much larger forces, have access to all of the state's resources and logistics, and affect almost the entire population.

So, why have we transitioned from wars between nations to civil wars? The video attributes this change to several factors, including the end of colonialism and the Cold War. The end of the Cold War removed one of the major drivers of armed conflicts and the break-up of communist dictatorships revealed new or old tensions, resulting in civil wars in the newly freed states. Moreover, in 1945, much of

Africa, Asia, and parts of Latin America were under colonial rule. However, by 1990, all but a few islands had gained independence. This newfound independence brought with it new conflicts, but it's important to note that colonialism was much worse than the current state of multinational corporations in the third world.

One of the key reasons for the decrease in wars between nations is democratization, the steady development from autocracy to democracy. Historically, democracies have rarely fought each other, and of all the state against state wars fought since 1900, only a minority were fought between democracies. Another reason is globalization, as war is no longer as effective as it used to be in achieving economic goals. It's now cheaper to buy resources on the global market than to seize them by force, and people from other nations are more valuable alive than dead, which is a relatively new concept.

War is now seen as a thing of the past, with rules that declare acts of aggression illegal and stipulate that armed forces are only justified in self-defence or with the authority of the UN Security Council. These rules are still broken, but it's now harder to do so without facing opposition and disapproval. Furthermore, there is now an international court for war crimes in The Hague, which is a recent innovation. Finally, borders are mostly fixed, with most countries pledging to accept international borders and respect other nations' autonomy after World War II.

It's too soon to say whether war is truly over, as we need a larger sample size to rule out the historical average of one or two big wars per century since World War II. However, if we don't have one major war in the next 75 years, it will be a strong indication that humanity is changing and moving towards lasting peace. This can be achieved by speaking up for peace and democracy and promoting these values in our daily lives.

In conclusion, while war and violence still exist in the world, it's important to recognize that we are in the most peaceful period in human history. The decrease in wars between nations can be attributed to several factors, including democratization, globalization, the changing perception of war, and fixed borders and international rules against aggression.

"The Solar Inferno: What Happens When the Sun Comes to Visit"

Bringing a piece of the Sun to Earth is not something one should ever attempt. The Sun is not solid, liquid or gas, but a hot ball of plasma. Imagine it as a goo-like substance, where the deeper one goes, the denser and weirder it becomes. So, let's consider bringing three samples of the Sun to Earth and see what happens.

Sample 1: Chromosphere

The Chromosphere is the atmosphere of the Sun, a layer of sparse gas that's covered in plasma spikes. It's pretty hot at 6,000 to 20,000 degrees Celsius, but if we brought a sample of it to Earth, it would immediately implode due to the Earth's atmospheric pressure. The vacuum created would cause air to rush in, releasing the energy of 12 kilograms of TNT, creating a high-pressure shockwave that could shatter glass, rupture ear drums and even kill someone close by.

Sample 2: Photosphere

Beneath the chromosphere is the photosphere, the glowing surface of the Sun that produces light. It's covered in a grid of hot spots called granules, each of which is over 5,000 degrees Celsius. If we brought a sample from here to Earth, it would instantly glow with a million times the brightness of the Sun, lighting up everything in the lab and then quickly cooling down to harmless gas.

Sample 3: Radiative Zone

The radiative zone is where the plasma is about two million degrees Celsius and tightly packed, creating a maze for energy in the form of photons. If we brought a sample from here to Earth, it would explode with the power of a thermonuclear weapon, destroying the lab and the surrounding city.

Sample 4: Core

The core is the central part of the Sun where a third of its mass is compressed and the temperature is 15 million degrees, enough to make helium and power the Sun through nuclear fusion. If we brought a sample of the core to Earth, it would cause an explosion with the equivalent of 4,000 megatons of TNT, or 1.3 km high cube of energy. The explosion would be so catastrophic that it could end human civilization. If there are any survivors, the dust blown into the atmosphere would cause a small ice age, potentially controlling human-caused climate change for a few decades.

In conclusion, bringing a piece of the Sun to Earth is not advisable. The consequences would be dire, potentially leading to the end of human civilization. The Sun is a massive ball of hot plasma, and we should be content to observe it from afar and enjoy its beauty and warmth.

Organic or O-Busted? Separating Fact from Fiction in the Healthy Food Industry"

Organic food has become a buzzword in recent years, with many people opting for it over conventional food due to its perceived health benefits and more natural and ethical production methods. But what does the term "organic" even mean? In general, organic food is produced without the use of GMO seeds, synthetic fertilizers, or synthetic pesticides. Instead, organic farmers use more traditional methods such as crop rotation and organic fertilizers like compost or manure.

But does eating organic food actually make us healthier? The truth is, there is no clear-cut answer. While some studies have found that organic food contains higher levels of antioxidants and vitamins, the evidence is mixed, and overall, the differences in nutritional value are small. Additionally, while organic food may contain less pesticide residue compared to conventional food, this doesn't necessarily mean it's safer or more natural, as organic pesticides can be just as toxic as synthetic ones.

When it comes to the environment, the impact of organic vs. conventional farming is not so clear-cut either. A 2017 meta-analysis found that while organic farming uses less energy, it has similar greenhouse gas emissions and requires more land to produce the same number of crops as conventional farming. Ecotoxicity, however, is where organic farming has a clear advantage. But the demand for organic food is constantly rising, which can lead to unsustainable production methods and even fraud, as it becomes more difficult to ensure that organic standards and regulations are upheld in the global trade of organic foods.

Organic food isn't just a production method, it's an ideology for many. People opt for organic food because it feels like the right thing to do, regardless of whether it's actually better for their health or the environment. In the end, the debate about organic vs. conventional food is a subjective one, and it ultimately comes down to personal values and beliefs. The most important thing is to ensure that we are eating a balanced and varied diet, regardless of whether our food is organic or not.

In conclusion, the question of whether organic food is truly better remains a highly debated topic in the world of nutrition and agriculture. On one hand, organic food is often grown without the use of harmful chemicals and pesticides, making it a healthier option for consumers. On the other hand, some studies have shown that the nutritional differences between organic and non-organic food are minimal, and the higher cost of organic products can be a barrier for many people. Additionally, the definition of "organic" can be misleading, and not all organic products are created equal.

Ultimately, the decision of whether to choose organic or non-organic food comes down to personal preference, health concerns, and financial situation. While there are many benefits to consuming organic food, it is important to educate oneself and make informed decisions based on individual needs and circumstances. The idea of "organic is better" should not be blindly accepted without considering the various factors that come into play.

"Drugs, A War We Can't Afford to Lose, But Can't Win Either"

The War on Drugs, initiated over 40 years ago by US President Richard Nixon, aimed to combat drug abuse and create a world without drugs. However, the outcome has been far from what was intended. The War on Drugs has resulted in disastrous consequences, both domestically and globally. The core strategy of the war, to eradicate the supply of drugs and incarcerate drug traffickers, has been a failure. By ignoring market forces, the war has led to a rise in the production of drugs and recruitment of traffickers, resulting in more drugs being available.

The drug market is not price-sensitive, meaning that drugs will be consumed no matter what they cost. This is known as the balloon effect. Even if production of drugs is stopped in one place, it will reappear in another. An example of this can be seen with crystal meth, where the US government tried to regulate the sale of chemicals used to produce the drug. This led to small-scale operations, but Mexican drug cartels eventually took over and opened large production operations, making the drug even more potent.

Despite a budget of around $30 billion and the efforts of the US Drug Enforcement Agency, the flow of drugs into the US and within the US is still widespread, with an efficiency rate of less than 1%. For many minors around the world, illegal drugs are as easy to access as alcohol.

Prohibition may prevent some people from taking drugs, but it also causes immense harm to society. Many of the problems associated with drug use are actually caused by the war against them. Prohibition makes drugs stronger, leading to increased consumption of more potent drugs. It also results in more violence and murders as gangs and cartels use violence to settle disputes. The incarceration of non-violent drug offenders is another devastating consequence of the War on Drugs. In the US, 25% of the world's prison population is incarcerated, largely due to harsh punishments and mandatory minimums. African Americans are disproportionately affected, with 40% of US prison inmates being black.

However, there is a way out of this mess. Switzerland's approach to dealing with a public health crisis related to heroin use in the 1980s provides a successful alternative. The country adopted a harm reduction strategy, opening free heroin maintenance centres. The results were a sharp drop in drug-related crime, a decrease in HIV infections, a 50% decrease in heroin overdoses, and a reduction in drug-related street sex work and crime.

After 40 years of the War on Drugs, it is time to revaluate our approach to drug abuse and consider alternative methods. The current war has led to a system that violates human rights, costs vast sums of money, and creates a lot of human misery, all in pursuit of an unattainable goal. It is time to look beyond the supply-side approach and adopt evidence-based methods that are cost-effective and actually work.

In conclusion, the war on drugs has been a massive failure on multiple fronts. Despite being waged for several decades and costing billions of dollars, it has not succeeded in reducing drug use or drug-related crime. Instead, it has contributed to mass incarceration, perpetuated racial disparities in the criminal justice system, and created new, unintended problems such as the opioid epidemic.

The "tough on crime" approach to drug policy has not only failed to solve the problem but has also created additional harm to communities and individuals. It's time for a new, more evidence-based and compassionate approach that prioritizes public health and human rights. By treating drug use as a health issue rather than a criminal justice issue, we can reduce harm to individuals and communities, and start to address the root causes of drug use and addiction.

"The Earth Strikes Back: A Crusade Against Climate Change"

The world is currently facing a severe crisis in the form of rapid climate change. Our planet is undergoing changes that are destabilizing our ecosystem and threatening the very survival of our civilization. Despite the urgency of the situation, most politicians appear to be unwilling to take meaningful action, and the fossil fuel industry remains actively opposed to change. As a result, many people have given up hope, feeling anxious, depressed and uncertain about the future.

However, the situation is not as hopeless as it seems. The last decade has seen a lot of positive developments that have accumulated and brought us closer to our climate goals. Despite the lack of action on the part of politicians and the influence of the fossil fuel industry, there has been a lot of progress made. In the last 20 years, the use of fossil fuels has declined, and the renewable energy sector has exploded with progress. Wind energy has become three times cheaper, and solar electricity is now ten times cheaper than traditional fossil fuels.

The story of the last decade is one of immense failure for climate policies around the world. Rather than passing comprehensive, binding bills that would reduce emissions, we have mostly done nothing. However, this story is not the whole picture. Despite the lack of action on the part of politicians and the continued lobbying and misinformation campaigns from the fossil fuel industries, there has been a lot of progress made.

Between 2000 and 2010, greenhouse gas emissions grew by 24%, driven by subsidies for fossil fuels aimed at promoting economic growth. In emerging countries such as China and India, coal was the cheapest fuel for growth, while rich countries showed little interest in changing their ways. However, the next decade turned out to be very different. The consumption of fossil fuels has decreased, and the growth of coal burning has slowed or levelled off in emerging countries and plummeted in rich countries. In the last decade, three-quarters of planned coal plants have been cancelled, and 44 countries have committed to stop building them.

Coal is no longer competitive, and technologies that were once expensive have matured rapidly. Renewable electricity has shown explosive progress, and wind and solar energy have become much more affordable. The decline of coal and the rise of renewable energy has had a significant impact on greenhouse gas emissions, and most scientists now believe that we have likely avoided catastrophic climate change. Although there is still substantial risk, we can now say with confidence that civilization will endure, even if it will have to change.

In conclusion, the situation may seem grim and hopeless, but the truth is that there is still a lot of reason to be optimistic. Despite the lack of action on the part of politicians and the influence of the fossil fuel industry, there has been a lot of progress made in the last decade. Renewable energy has become more affordable, and the use of fossil fuels has declined. Most scientists now believe that we have likely avoided catastrophic climate change, and that civilization will endure, even if it will have to change. We must not give up hope, and instead continue to work towards a better future for our planet and ourselves.

"From Tiny to Titans: The Limits of Bacterial Growth"

The journey of life is about solving problems and staying alive, and one of the most important factors that determines how these problems are solved is size. The physical laws of the universe, such as temperature, microgravity, and surface tension, can have different consequences for living things depending on their size.

For living things to survive, they need to transport essential materials from the outside to the inside. At the beginning of life on Earth, this was a challenge, as energy was required to transport materials and the first living beings did not have the abundance of tools and techniques available today. However, they found a solution in the form of diffusion.

Diffusion is a physical law that states that molecules in liquids or gases are constantly moving around and bumping into each other, spreading out evenly over time. As it doesn't require energy, life on Earth relies on diffusion to transport materials. The smallest living thing on Earth, a bacterium, uses diffusion to transport oxygen and carbon dioxide through its cell membrane. This "breathing" mechanism is only effective for very small creatures, such as bacteria and cells, and insects with a fine network of trachea.

The fundamental problem is that as living things get larger, diffusion becomes too slow to keep them alive. This is because the exchange of materials with the environment can only happen at the surface and the amount of material that can be transported is limited. The square-cube law states that as an object grows larger, its surface area grows at a slower rate than its volume. This means that a larger living thing would have too much volume and most of it would be far from its surface, making it impossible for essential materials to reach its interior.

To overcome this problem, life evolved into multicellular structures composed of many cells instead of one. This way, diffusion works better as it can happen in each individual cell. Over time, cells began to share work and specialize, with some cells concentrating on sensing the environment, others on digestion, and others on movement. To become even larger, life solved the diffusion problem by having holes, caves, tunnels, and by folding in on itself, so that diffusion could happen easily in each cell.

For example, a human's skin has a surface area of about 2 square meters, while the lungs have a surface area of about 70 square meters. They can expand and contract to increase their surface area, making it easier for oxygen to diffuse into the body. The circulatory system was also developed, which acts as a pipeline to transport essential materials to cells that are far from the surface.

In conclusion, size is a crucial factor that determines how living things solve problems and stay alive. The smaller the living thing, the easier it is to transport essential materials through diffusion. However, as living things grow larger, they face problems with diffusion and need to evolve new ways to transport essential materials. Multicellular structures and specialized systems, such as the circulatory system, have allowed life to overcome these problems and become larger.

"Mining the Heavens: The Quest for Unlimited Resources in Space"

Asteroid mining, a concept that has been long speculated, is finally becoming a reality with advancements in space travel. The idea of extracting valuable resources from asteroids has been around since the dawn of space exploration, but it wasn't until recent years that the concept became economically feasible.

Asteroids are millions of tons of rocks, metals, and ice left over from the formation of the solar system over 4.5 billion years ago. Most of these celestial bodies are found in the asteroid belt and the Kuiper belt, with hundreds of thousands more orbiting independently between the planets. These asteroids are not only rich in industrial and precious metals, but they also hold the key to solving some of the world's most pressing problems.

The mining industry on Earth is plagued by environmental degradation and the use of dangerous chemicals that harm both people and the planet. The industry is also subject to geopolitical manipulation as countries restrict access to valuable resources. Asteroid mining offers a clean and sustainable alternative to these problems by providing unlimited resources from space.

Asteroids contain vast quantities of precious metals and industrial materials, such as platinum, gold, and iron. For example, the 16 Psyche asteroid alone contains enough iron nickel to cover the world's metal needs for millions of years. While the concept of asteroid mining sounds promising, the reality of extracting resources from space is much more complicated.

The main challenge facing the industry is the cost of space travel, which currently stands at thousands of dollars per kilogram of payload.

In order to make asteroid mining economically feasible, space travel must be made cheaper and more efficient. One solution to this problem is to switch from traditional rockets to electric spaceships. Electric engines are not powerful enough to launch a spaceship into orbit, but they are efficient once they are in space, making space travel much more affordable.

Once a spacecraft has arrived at an asteroid, the next step is to secure it and stop it from spinning. This can be achieved by vaporizing material with a laser or using thrusters to stop the rotation. After the asteroid has been stabilized, it is moved into a trajectory that takes it near the Moon, where its gravitational pull can be used to put the asteroid in a stable orbit around Earth, saving even more fuel.

The next step is to process the asteroid. The process begins with heating up the rock with sunlight, which boils out the gases. The rock is then ground into gravel and dust, and the dense and light elements are separated using centrifuges. Although the extraction process is expensive, it is still more cost-effective than traditional mining on Earth, as even extracting just 0.01% of an asteroid's mass can yield several times more valuable resources than an equivalent amount of ore on the ground.

Finally, the precious metals must be transported back to Earth. The materials can be loaded into reusable rockets or 3-D printed heat-shielded capsules filled with gas bubbles, which can be dropped into the oceans and towed away by ships.

Asteroid mining represents a new frontier for humanity, and it has the potential to revolutionize the way we extract resources and build infrastructure. As our experience and technology continue to grow, so too will our missions become more sophisticated and self-sustaining. Parts and fuel produced on asteroids won't have to be launched from Earth, making subsequent missions easier and more efficient.

In conclusion, asteroid mining offers a clean and sustainable alternative to traditional mining on Earth. With advancements in space travel, it is now possible to extract unlimited resources from space, providing humanity with the materials it needs to build a better future. By embracing this new frontier, we can begin our first real steps towards colonizing the solar system and creating a better world for future generations.

"The Price of Progress: The True Cost of each Energy Source"

Nuclear energy has been a source of electricity for over 70 years and in that time, there have been about 30 reported accidents globally. Although most of these accidents were minor, two stand out as the worst nuclear disasters in history: Chernobyl and Fukushima.

Chernobyl is considered the worst disaster due to the age of the reactor technology and the government's slow response. Only 31 people died directly in the accident, but the radiation released from the explosion is the real danger, causing cancer and other diseases. Estimations of the death toll caused by the radiation range from 4,000 according to the WHO to 60,000 according to a study commissioned by the European Green Party.

Fukushima was a much better prepared accident, with more advanced technology and security measures in place. The death toll stands at 573, but these deaths were not a result of radiation, but rather from the stress of the evacuation of the surrounding area. The long-term death toll from radiation is estimated to be anywhere from none to 1,000.

Renewable energy sources, such as solar, wind, and geothermal, cause minimal deaths due to accidents during construction and maintenance. Hydropower, the most widely used renewable energy source, has caused the most fatalities, with hundreds of thousands of deaths in the past 50 years. One of the deadliest accidents was the 1975 Banqiao hydroelectric dam failure in China, which was caused by a massive typhoon, poor design, and poor management. The death toll from this one accident is estimated to be between 85,000 to 240,000.

Fossil fuels, however, are the real killer energy source. When burned, they release gases such as ozone, sulphur dioxide, carbon monoxide, and nitrogen dioxide, which cause a wide range of respiratory and cardiovascular diseases, as well as lung cancer and other deadly conditions. The damage caused by air pollution from fossil fuels happens gradually, making it difficult for people to realize the scope of the problem. Air pollution from fossil fuels is estimated to cause the deaths of 4 million people each year and has killed around 100 million people in the past 50 years.

In conclusion, while nuclear energy has caused the worst disasters in history, it still pales in comparison to the number of deaths caused by fossil fuels. Renewable energy sources cause the least number of deaths, but hydropower has had the most fatalities. The true cost of each energy source must be weighed in terms of not only the number of deaths caused but also the amount of energy produced.

"From Pisa to Wall Street: How Banking Evolved Over Time"

Banking is a complex and mysterious system that holds an immense amount of assets globally. With over 30,000 different banks worldwide, it can be challenging to understand how they work. But it all started with a simple idea - to make life easier. In the 11th century, Italy was the centre of European trading, and merchants from all over the continent came to trade goods. The problem was that there were too many currencies in circulation, and merchants had to deal with multiple types of coins, exchanging them constantly. This exchange business, which took place on outdoor benches, gave birth to the word "bank."

As time passed, the dangers of traveling, counterfeit money, and the difficulty of obtaining a loan prompted the creation of a new business model. Home brokers started to give credit to businessmen, while Genevese merchants developed cashless payments. Banks spread all over Europe, providing credit even to the church and European kings.

Today, banks are in the risk management business. People keep their money in banks, receiving a small amount of interest, and the bank takes that money and lends it out at much higher interest rates. It's a calculated risk, as some lenders will default on their credit. This process is crucial for our economy, as it provides resources for people to buy houses and for industries to expand their businesses. Banks turn unused funds from savers into funds society can use for growth.

Other sources of income for banks include accepting saving deposits, the credit card business, buying and selling currencies, custodian business, and cash management services. The main problem with banks today is that many have abandoned their traditional role as providers of long-term financial products in favour of short-term gains with much higher risks. During the financial boom, most major banks adopted incomprehensible financial constructs, which ultimately led to the global banking crisis of 2008. This crisis caused the collapse of the housing market in the US and parts of Europe, leading to the loss of millions of jobs and billions of dollars.

As a result, new regulations were put into place to govern the banking business, but some tough legislation was successfully blocked by the banking lobby. Alternatives to traditional banking are gaining popularity, such as investment banks that charge yearly fees and do not earn commissions on sales. Credit unions, cooperative initiatives established in the 19th century, provide the same financial services as banks but focus on shared value rather than profit maximization. Crowdfunding has also exploded in recent years, providing loans from small investors, avoiding the bank as a middleman. Micro-credits, small loans mostly handed out in developing countries, help people escape poverty.

In conclusion, banking plays a crucial role in society by providing funds to people and businesses, but the way it is done and who does it will continue to evolve. It is up to us to decide the future of banking.

"The African Opportunity: Addressing Overpopulation for a Better Tomorrow"

The world's population has experienced a rapid growth in the past century, with the number of people on earth increasing fourfold in a span of a hundred years. This has sparked concerns about overpopulation and its effects on the earth. But population growth rate has peaked in the 1960s and has since slowed down as countries industrialize and develop. Currently, the world's population is estimated to reach 11 billion by the end of the century.

However, the global picture masks the details of population growth in different regions of the world. One such region is Sub-Saharan Africa, which is home to a billion people living in 46 countries. Although the growth rate in this region has slowed in recent decades, it is still much higher compared to the rest of the world. Projections estimate that the population of Sub-Saharan Africa could reach anywhere between 2.6 billion to 5 billion people by 2100.

This growth presents a huge challenge for any society, especially for Sub-Saharan Africa, which is the poorest region on earth. So, is Sub-Saharan Africa doomed to suffer from overpopulation and poverty? The answer is not that simple, as there are many factors that influence population growth and poverty in this region.

For example, many countries in Asia were at a similar stage as Sub-Saharan Africa today a few decades ago, with large populations living in extreme poverty and high birth rates. Countries like Bangladesh have successfully implemented family planning programs that have greatly reduced population growth, child mortality, and improved the country's economy.

Similarly, investment in health and education has led to lower birth rates in other Asian countries like South Korea, India, Thailand, and the Philippines, which has resulted in improved demographics and economic growth. The same progress has not been seen in Sub-Saharan Africa, primarily due to a variety of reasons, including colonization, civil wars, unstable governments, cultural differences, and insufficient aid.

For example, in Sub-Saharan Africa, education has not improved as rapidly as in other parts of the world, and the unmet need for modern contraception is still at 60% among adolescents. The reasons for these challenges are diverse and complex and cannot be summarized in a few sentences.

If population growth in Sub-Saharan Africa continues at its present rate, it could reach over 4 billion people by 2100, presenting a massive challenge for the region. But there is still hope, and much can be done to address this issue. Investment and aid that focuses on building systems for education, family planning, and health care can greatly impact the region's population growth and improve the quality of life for its people.

For example, if women receive better education and have access to family planning resources, it could result in lower birth rates, better health outcomes, and improved economic growth. Other initiatives, such as improving healthcare and providing access to contraceptives, can also have a significant impact.

In conclusion, while overpopulation is a global concern, it presents unique challenges in Sub-Saharan Africa. However, with investment, aid, and the right initiatives, it is possible to address this issue and improve the lives of the people living in this region.

"Choking on Plastic: Our Dangerous Dependence"

Plastic, the magic material, has saturated our environment and is invading our food chain. Plastic is a synthetic polymer made from crude oil components, which have extraordinary traits - lightweight, durable, and mouldable. However, these same properties make plastic incredibly difficult to break down, taking between 500-1000 years. Despite this, plastic has become ubiquitous in our lives, and we use it for almost everything, from clothes to houses and cars. In the past 100 years, we have produced 8.3 billion metric tons of plastic, with over 6.3 billion metric tons becoming waste. Unfortunately, only 9% of this waste was recycled, 12% burned, and 79% remains stuck in our environment, much of it ending up in our oceans. This plastic pollution is having a devastating impact on marine life, with over 90% of seabirds having already ingested plastic, and many animals dying with stomachs full of indigestible trash.

Microplastics, particles smaller than 5 millimetres, are an even more widespread and invisible form of plastic pollution. These microplastics result from floating waste that crumbles into smaller and smaller pieces, and there are now an estimated 51 trillion particles floating in our oceans. These microplastics are easily ingested by marine life and are traveling up the food chain, with evidence of microplastics being found in our tap water, household dust, and even in our bodies. There is little scientific evidence about the health risks posed by microplastics, but it is known that many plastic additives, such as BPA and DEHP, can interfere with our hormonal systems and may even cause cancer.

Banning plastic may seem like the solution to plastic pollution, but it's not that simple. Some of the alternatives to plastic have a higher environmental impact in other ways and finding the right balance between trade-offs is a complex process. Plastic also solves problems that we don't have good answers to yet, and so a blanket ban may not be the best solution. Instead, we need to find ways to reduce our plastic usage, recycle more effectively, and find alternative materials that are both environmentally friendly and sustainable.

In conclusion, plastic has become a double-edged sword. It has revolutionized our lives, but the convenience it provides has come at a steep cost. Plastic pollution is a global crisis, and it's up to each and every one of us to play our part in reducing our plastic footprint and ensuring a healthy planet for future generations.

"The Messenger of Information: Decoding the Nature of Light"

Light is a magical phenomenon that has captivated humans for centuries. It is the connection between us and the universe, allowing us to see and experience the world around us. It is the smallest unit of energy that can be transported and is both a particle and a wave simultaneously. Light is part of the electromagnetic spectrum, with visible light being only a small portion of it. The wavelengths of visible light range from 400 to 700 nanometres, and it is the only set of electromagnetic radiation that propagates in water, where eyes first evolved millions of years ago.

Light is created when atoms or molecules drop from a higher energy state to a lower one and emit energy in the form of radiation. At the microscopic level, light is created when an electron within an atom drops to a lower energy state and loses excess energy. This movement of electrons creates oscillating electric and magnetic fields, which transfer energy and carry information about the place of creation.

Light travels at the speed of light, which is 299,792,458 meters per second in a vacuum, and is propelled by oscillating electric and magnetic fields. Any particle without mass travels at the speed of light and it is the fastest way to travel through space in the universe. The light that is released from a candle does not speed up until it reaches light speed, and at the very moment of its creation, its speed is c.

The reason why light is finite and travels at its maximum speed is still unknown. There is much more to learn about the properties of light, including the complex topics of time and the twin paradox. However, we can be grateful that we have evolved eyes that pick up the waves of information permeating the universe, allowing us to see things and put our existence into perspective.

Light is not just a mere physical phenomenon, it is also a symbol of hope, inspiration, and wonder. It illuminates the darkness and reveals the beauty of the world. It dances in the sky during a sunset and twinkles in the night sky as stars. It sparkles on the surface of a calm lake and reflects off the snow-covered mountains. Light is a reminder that there is more to this world than what we can see and touch, it is a glimpse into the infinite beauty of the universe.

In conclusion, light is a mysterious and magnificent force that continues to amaze and inspire us. It is the connection between us and the universe, allowing us to see and experience the world around us. It is a reminder that there is so much more to this world than meets the eye and that there is always something new to discover. Light is a symbol of hope, inspiration, and wonder, and it will continue to be a source of awe and wonder for generations to come.

"The Timeless Debate: Should We End Aging?"

The concept of aging has been a subject of fascination for human beings for thousands of years. It's only natural to wonder about our lifespan and the inevitable process of aging that we all must face. But what if aging wasn't inevitable? What if we had the power to stop it? This is the exciting possibility that recent advancements in longevity research have opened for us.

Aging is caused by physical processes like oxygen, radiation from the sun, and metabolism that wear down our bodies over time. While our bodies have mechanisms to repair this damage, they become less effective as we age, leading to weakened bones, wrinkled skin, a declining immune system, and the onset of diseases that ultimately lead to death. But what if we could stop these processes from happening?

Longevity research has made unprecedented advances in recent years and scientists are now starting to understand the mechanisms behind aging and how to manipulate them. Aging is not mystical or inevitable, and with the right tools, it might be possible to delay or even stop it. This opens a world of possibilities, including the end of disease, the prolongation of healthy lifespans, and the ability to live life to its fullest for longer.

But should we end aging? The idea of life extension makes many people uncomfortable, as it goes against the natural order of things. Humans have always been born, grown old, and then died, and the idea of avoiding death entirely is unsettling for some. But when you consider what life extension really means, it's just another phrase for medicine. Healthcare professionals are just trying to prolong life and minimize suffering, and the vast majority of resources are spent on the consequences of aging.

By trying to stop aging from happening, we would be doing something that is no less natural than transplants, chemotherapy, antibiotics, or vaccines. Nothing humans do nowadays is purely natural anymore, and we enjoy the highest standard of living ever as a result. Right now, we wait until it's too late and the machine is failing, then use resources to try and fix it, while it continues to break down further.

Ending aging would not be the end of death, but rather like a summer evening when you were a kid, and your mom called you inside. You just wanted to keep playing for a little longer, until you felt tired. Imagine a world without disease where you and your loved ones could live in good health for another 100 or 200 years. How would that change us? Would we take better care of our planet if we knew we would be around longer? Would we spend more time learning and figuring out what we're good at?

The possibilities are endless, and the end of aging would bring about a new era in human history. It would mean longer, healthier lives, the ability to live life to its fullest, and a world where disease is a thing of the past. This is the future that science and technology are bringing us closer to every day, and it's up to us to decide whether or not we want to embrace it.

"A Universe in Peril: Three Paths to Oblivion"

One day, the universe will come to an end, but the question remains, how will it happen and what will happen after? Scientists believe that the universe is expanding, and that this expansion is accelerating due to the strange phenomenon of dark energy. Despite extensive research, the properties of dark energy are still unknown, but there are various theories that suggest three scenarios for the end of the universe: the Big Rip, Heat Death or Big Freeze, and Big Crunch and Big Bounce.

The first scenario is the Big Rip. The universe has been expanding since its birth, and over time, the expansion accelerates to a point where space expands so fast that gravity cannot compensate, leading to the Big Rip. Galaxies, black holes, stars, and planets would all dissolve into their components, and in the end, atoms would disband. Space would be expanding faster than the speed of light, and no particle in the universe could interact with any other particle, leading to a strange and timeless universe.

The second scenario is Heat Death or Big Freeze. Every system tends towards the state of highest entropy, and this also applies to the universe. As the universe expands, matter decays and spreads out until all the gas clouds necessary to form stars are exhausted, and the universe turns dark. The remaining suns die, black holes slowly degrade and evaporate, and eventually, only a dilute gas of photons and light particles remains. All activity in the universe would cease, entropy is at its maximum, and the universe would be dead forever. However, it is theoretically possible that after an incredibly long amount of time, there might be a spontaneous entropy decrease leading to a new Big Bang.

The third and final scenario is the Big Crunch and Big Bounce. If there is less dark energy than thought, or it decreases over time, gravity will be the dominating force in the universe one day. The rate of expansion will slow down and stop, then reverse, and galaxies will race towards each other, merging as the universe becomes smaller and smaller. The universe would become hot, and background radiation would be hotter than the surfaces of the most stars. Atoms would be ripped apart, before supermassive black holes devour everything, leaving a supermassive mega-black hole that contains the entire mass of the universe. The Big Bounce theory states that this has happened many times and that the universe goes through an infinite cycle of expansion and contraction.

At the moment, heat death seems the most likely scenario, but scientists hope that this "dead forever" idea is incorrect, and the universe will start over and over again. However, no one knows for sure, but the most uplifting theory is always a good assumption. The end of the universe is a fascinating topic, and scientists continue to study it, hoping to shed light on what will happen and what the universe will become. The universe is a vast and mysterious place, and there is still so much to discover and understand. Who knows what other theories and scenarios will emerge in the future? The universe is waiting to be explored and understood, and the journey is just beginning.

"The Cosmic Loopholes: Understanding the Mysteries of Wormholes"

Wormholes have long been a fascinating topic of discussion in the world of science fiction, but could they actually exist in our universe? The concept of wormholes dates back to Einstein's theory of relativity, which states that space and time are not constant and can be affected by the presence of matter. This idea of elastic space has led scientists to believe that wormholes, essentially shortcuts through spacetime, might be possible.

A wormhole is a theoretical passage through space and time that connects two distant points in the universe. It's often depicted as a spherical or circular object, with light from the other side shining through, providing a window to a faraway place. Once you cross a wormhole, you'd find yourself in a completely different part of the universe, with your previous location now just a shimmering window.

The first type of wormhole theorized was the Einstein Rosen Bridge, which describes black holes as portals to parallel universes. According to this theory, a black hole's event horizon is a one-way barrier that anything can enter but nothing can escape. It's possible that beyond the event horizon lies a parallel universe with time running in reverse. However, crossing an Einstein Rosen Bridge would take an infinite amount of time and is impossible in reality.

To travel through the cosmos, humans would need a traversable wormhole, a type of wormhole that is possible to cross. If string theory is correct, there could already be countless traversable wormholes scattered throughout the universe, created in the earliest moments of the Big Bang by quantum fluctuations at the smallest scales. Some physicists even suggest that the supermassive black holes in the centres of galaxies might be wormholes.

Creating a wormhole is a challenging task, however. First, it must connect distant parts of space-time, like Earth and Jupiter. Second, it should not have any event horizons that would block two-way travel. Third, it must be big enough to not pose a threat to human travellers. The biggest challenge, however, is keeping the wormhole open. Gravity is always trying to pinch it closed, leaving only black holes at the ends. To keep the wormhole open, we need something called exotic matter, a new and exciting substance that is different from anything we have on Earth or even antimatter.

In conclusion, while wormholes are still just a theoretical concept, they have captured the imaginations of scientists and science fiction enthusiasts alike. The idea that we could one day travel through shortcuts in spacetime, connecting distant parts of the universe, is a thrilling prospect. Although we are still far from making it a reality, the possibility of wormholes is a testament to the incredible power of human imagination and the limitless possibilities of the universe we inhabit.

"Nightmares Come Alive: The Brain-Eating Amoeba's Rampage"

The world is full of billions of microscopic creatures, most of which pose no harm to humans. However, there is one exception to this rule: Naegleria Fowleri. This brain-eating amoeba is a microbe with a nucleus and is one of the smallest life forms on earth. It is a voracious hunter, devouring bacteria and other critters whole, and is often found in freshwater bodies like ponds, rivers, lakes, and hot springs. In the summer, when water temperatures are high, it thrives and multiplies, making it difficult for humans to avoid contact with the amoeba.

When people dive or swim in water contaminated with Naegleria Fowleri, and water splashes high into their noses, it can lead to disastrous consequences. This amoeba is particularly good at evading the human immune system, as it is able to fly under the radar of the mucosa, a slime filled with chemicals that normally kill or stun invaders. The inside of the nose is filled with a large network of olfactory nerve cells that transmit information to the olfactory bulb, the centre of the sense of smell in the brain. Unfortunately, these nerve cells release a chemical called acetylcholine, which Naegleria Fowleri is irresistibly attracted to.

As the olfactory nerve cells release acetylcholine, Naegleria Fowleri enters the tissue, following the chemical signals upstream. The immune system's suicide warriors, known as neutrophils, attack the amoebae, but they are no match for their size and strength. Despite the attacks, Naegleria Fowleri continues to follow the olfactory nerves to its final destination: the human brain. The process can take anywhere from one to nine days, during which time the person is unlikely to notice anything.

Once the amoebae reach the olfactory bulb, the entrance to the brain, they initiate a massacre, releasing an onslaught of attack molecules. The brain cells become helpless victims, and Naegleria Fowleri multiplies and begins to feed on them. In a feeding frenzy, the amoeba can develop up to a dozen food cups, which look like giant eerie mouths. It engages the brain cells, sucking them in and ripping large pieces out of them while they are still alive. The disease sets in quickly, and death is usually inevitable.

In conclusion, Naegleria Fowleri is a monster in the world of microscopic creatures, with a deadly taste for human brains. While millions of people have regular contact with the amoeba, it is important to be aware of the dangers it poses, especially in warmer climates during the summer months. To protect oneself from the brain-eating amoeba, it is advisable to avoid diving or swimming in contaminated water, and to take precautions to ensure that water is properly treated, especially in pipes, swimming pools, fountains, or spas.

"The Miniature Masterpiece: The Science of Atoms"

Atoms, the building blocks of all matter in the universe, are incredibly small and nearly impossible to imagine. The thickness of a single human hair is about as thick as 500,000 carbon atoms stacked on top of each other. If you think about a single atom in your fist as big as a marble, your fist would be the size of the Earth.

An atom is made up of three fundamental particles: neutrons, protons, and electrons. The protons and neutrons form the atom's core and are held together by the strong interaction, one of the four fundamental forces in the universe. This core is made up of even smaller particles called quarks, which are held together by gluons. Quarks and electrons are believed to be the most fundamental components of matter in the universe, but the exact size of quarks is unknown, and they may be zero-dimensional points.

Electrons orbit the atom's core and travel at a speed of about 2,200 km/s, which is fast enough to get around the Earth in just over 18 seconds. Despite the fast speed of electrons, 99.9999999999999% of an atom's volume is empty space filled by quantum fluctuations, which are fields with potential energy that build and dissolve spontaneously. These fluctuations have a significant impact on how charged particles interact with each other.

If you were to remove all the spaces between the atom cores from the Empire State Building, it would be about the size of a grain of rice. All the atoms of humanity would fit in a teaspoon. In neutron stars, atoms are compressed so densely that the mass of three suns fits into an object only a few kilometres wide.

The current understanding of atoms is that electrons are like a wave function and a particle at the same time. Scientists can calculate where an electron might be at any given moment, and these clouds of probability are called orbitals. The probability of finding an electron approaches 0 the further away from the atom core, but it never reaches zero, which means that in theory, an electron could be on the other side of the universe.

For many elements, only three fundamental particles are needed to form different atoms. Hydrogen is made up of one proton and one electron, while helium is made up of a proton and a neutron. The addition of more particles creates elements such as carbon, fluorine, and gold. Every atom of an element is the same, regardless of where it is in the universe.

The understanding of atoms has changed multiple times throughout history, and it is likely that this current model will not be the last. Scientists continue to research and discover new information about the strange world of atoms, which is the basis for our existence. The study of atoms and their behaviour is a complex and fascinating subject, and it is exciting to think about what new information will be uncovered in the future.

"A Giant Leap for Mankind: The Space Elevator and Beyond"

The thought of exploring the vastness of space has captured the imaginations of humans for centuries. But currently, the only way to experience space is to either become an astronaut or have a billionaire's bank account. However, there is a concept that may make it possible for the average person to experience space and even become a space-faring civilization – the space elevator.

The space elevator is essentially a long tether that extends from the Earth's surface to space, allowing a climber, much like an elevator carriage, to move up and down the tether. The tether is anchored to the Earth and held tight by a counterweight located at the top, higher than 36,000 kilometres above the Earth's surface. The counterweight holds up the tether, and the tension from the counterweight supports the tether, much like a rope. At the counterweight, there could be a space station, which serves as a launching point for all missions from the spaceport elevator.

The concept of a space elevator is based on the principles of orbits. To enter Earth's orbit, rockets have to move up and sideways fast, but with a space elevator, the energy required to go up is provided, and the fast sideways movement comes for free with Earth's rotation. This is compared to rockets, which burn a huge amount of rocket fuel just to get a small amount of cargo into space, making it extremely expensive. The cost of putting one kilogram of payload into space is currently around $20,000, making human spaceflight one of the biggest limitations due to its immense cost. A space elevator is projected to reduce the cost one hundredfold, to $200 per kilogram, making it a cost-effective solution.

However, building a space elevator is not without its challenges. The biggest challenge is the tether, which needs to be light, affordable, and more stable than any material currently available. Some promising materials like graphene and diamond nano threads have been proposed, but they may not be strong enough to withstand the rigors of space. The tether must also be able to withstand atmospheric corrosion, radiation, and micrometeorite and debris impacts. Additionally, the climber requires a lot of energy to go up, and it is still unknown how to power the climber. Some experts have proposed using a nuclear reactor on the elevator carriage, while others have suggested beaming power from the ground with a super-powered laser. The raw materials for a 36,000-kilometer-long tether must also be obtained, either by making it on Earth and launching it into space or by making it in space and lowering it down to Earth.

Aside from the technological challenges, there is also the risk of the tether breaking. If it breaks near the anchor, the force exerted by the counterweight will cause the entire elevator to rise up into space. If it breaks near the counterweight, the tether will fall, wrapping around the world and whipping the end off, creating debris in orbit that could pose serious problems for future spaceflight. Some experts have even proposed building a space elevator on the Moon, as the Moon's gravity is much weaker and a flimsy material like Kevlar could serve as a tether.

Despite the challenges, the payoff of having a working space elevator would be immense. It could be the first step to becoming a space-faring civilization and open up the doors to the exploration of the universe. Even if a space elevator is never built, in trying to build one, we may learn an awful lot about space, materials, and technology. The idea of a space elevator is a dream of the future, and one that has the potential to change the course of human history.

In conclusion, the concept of a space elevator presents a ground-breaking opportunity for mankind to reach space in a more affordable and efficient manner. Despite the technological and scientific hurdles that still need to be overcome, such as the development of a suitable tether material and the resolution of power and transportation issues, the potential benefits of a space elevator are numerous. It has the potential to reduce the cost of getting payloads into space by 100 times, making it possible for more individuals and organizations to access space. The development of a space elevator could also serve as a stepping stone towards becoming a space-faring civilization, with the ability to launch missions from a space station at the top of the elevator. However, the risks associated with building a space elevator must also be taken into consideration. A tether break, either near the anchor or the counterweight, could result in catastrophic consequences, making it imperative to get it right the first time. Despite these challenges, the pursuit of a space elevator is a dream worth pursuing as it has the potential to bring about a glorious future for space exploration and discovery.

The Space Elevator: A Dream of the Future

"Finding Home in the Unknown: The Power of Perception in Location"

It's a common misconception that our position in the universe is absolute and that the world we occupy is a static and unchanging stage beneath our feet. However, the truth is that our perception of position is relative and created by humans. The universe is a vast expanse of space, and if we were to remove all the things within it, there would be nothing left but empty space, uniform in every direction. In empty space, the concept of position becomes meaningless.

Our perception of the world is limited by our frame of reference, the way we see the universe and how we perceive things moving around us. From our perspective, the world appears flat and we can move in three dimensions, but this is just one frame of reference. In reality, the ground curves away from us after only five kilometres from our location, and if we could see through it, we would see people from below or sideways. But they don't fall off the planet because the concept of "down" is an illusion of our frame of reference. In the frame of reference of the Earth, gravity just pulls inward.

The Earth is constantly moving, never staying in one position. It orbits the sun, which is at the centre of the solar system, and the sun is moving too. The orbit of the Earth is not a perfect circle but an ellipse that changes shape every 100,000 years, and the orbit itself drifts in a cycle of 112,000 years. The moon also affects Earth's movement, causing it to jiggle as it orbits both objects around their common centre of gravity.

From the perspective of the sun, the plane of the solar system is arbitrary, defined as the plane the Earth orbits in because it is convenient for us. In reality, the other planets are just a little inclined with respect to our plane, and we are the ones with a slightly bent orbit. The solar system as a whole orbits the centre of the Milky Way galaxy, but the plane of the solar system is not aligned with the plane of the galaxy. All the stars orbiting the galactic centre have their own planes, and the solar system as a whole is tilted about 60 degrees towards the galactic plane, moving through space at almost a million kilometres per hour.

From the centre of the galaxy, the orbits of the planets would appear to be moving in a helix shape, like a corkscrew motion, on the tilted plane of the solar system relative to the plane of the galaxy. This orientation in space means that sometimes the planets are in front of the sun, and sometimes they are behind it, always in a state of constant motion.

In conclusion, the concept of absolute position is a human construct, and our perception of the world is limited by our frame of reference. The Earth is constantly moving, orbiting the sun in an elliptical path that changes shape every 100,000 years, while the solar system orbits the centre of the Milky Way galaxy. Our position in the universe is relative, and we are always in a state of constant motion, moving through space on a helix-shaped path.

"The Electromagnetic Menace: Navigating the Dangers of Modern Technology"

Electricity is a ubiquitous part of our lives and has made it easier, safer and more enjoyable. But with the ever-increasing use of technology, the question arises, could too much electricity be hurting us? Could the foundation of our modern world slowly be killing us?

Electricity is the movement of an electric charge that generates electric and magnetic fields, known as Electromagnetic Radiation. The electromagnetic spectrum consists of various types of radiation, some of which can be dangerous, like UV light, X-rays, and Gamma rays. Other forms of radiation are harmless and include visible light, infrared, microwaves, and radio waves, which are emitted by human technology such as mobile phones, Wi-Fi routers, electric power lines, and household appliances.

This weak electromagnetic radiation doesn't disrupt the molecules in our bodies but can stimulate muscles and nerves, causing a tingling sensation. It can also heat up the water molecules in our food, making it warm. Despite being surrounded by natural and generally harmless sources of electromagnetic radiation, since the industrial revolution, we have added a lot to our immediate environment.

The idea of the potential danger of electromagnetic radiation gained public attention in 1979 when a study linked leukaemia to living near power lines. However, this study was quickly discredited, and no direct causal link was confirmed. Despite that, the idea persisted, and many studies have tried to establish the possible dangers of electromagnetic radiation. Some people claim to suffer from symptoms like headaches, nausea, skin reactions, burning eyes, or exhaustion due to exposure to this radiation, but these are just everyday effects.

There have been a few studies that have found more unsettling results, such as possible connections between the side of the brain people use when on their phone and the appearance of brain tumours. The question that science is trying to answer is not about the acute effects of irradiation, but whether weak electromagnetic radiation that we are constantly surrounded by is harmful in the long run due to some unknown mechanism.

However, the answer to this question has proven to be much harder than expected, due to the sheer number of studies, reports, and statements by various organizations. Many of the much-cited studies that spread panic about electromagnetic radiation are highly controversial, and often based on unreliable surveys and self-reporting. People tend to misremember things, and the media can easily influence public opinion by cherry-picking the findings that best suit their opinion or make for the most exciting headline.

For example, a study that looked for cancer in rats and mice from cell phone radiation seemed to show a connection, but only in male rats and not in mice. Despite that, it was reported as if it proved that mobile phone radiation causes cancer. The World Health Organization has officially classified radio frequency fields as possibly carcinogenic, meaning there are hints that they might cause cancer, but it cannot be proven, and the situation will be monitored.

On the whole, there was no consistent evidence in human studies that electromagnetic radiation below exposure value limits causes health problems. Although there are some statistical associations, they are not considered causal relationships. In conclusion, while the potential dangers of electromagnetic radiation cannot be completely ruled out, there is currently no consistent evidence that supports the notion that it is harmful to human health. Until more conclusive evidence becomes available, it is recommended to limit exposure to electromagnetic radiation, especially for children and pregnant women, as a precautionary measure.

"The Dark Celestial Mystery: The Black Hole Star"

Black Hole Stars - The Cosmic Wonders That Shouldn't Exist

Imagine a star so massive that it burns brighter than galaxies and dwarfs any star we know today. Now imagine this star is home to a cosmic parasite, a ravenous black hole that devours billions of tons of matter every second. This is the world of Black Hole Stars, a phenomenon that defies all we know about star formation and growth.

In the early universe, a few hundred million years after the Big Bang, the universe was much smaller and denser, dominated by dark matter and dark matter halos. These massive structures pulled in and concentrated massive amounts of hydrogen gas, leading to the formation of the first stars and galaxies. In this unique environment, the massive gravitational pull of the dark matter halos caused clouds of hydrogen to form that were 100 million times the mass of our sun.

This gave birth to the Black Hole Stars, young stars that were forced to grow and grow until they reached up to ten million times the mass of our sun. With a core that became hotter and hotter, the balance between gravity pulling in and radiation pushing out became impossible to uphold. The core was eventually crushed into a black hole, and although the star survived its own death, the explosion wasn't enough to destroy it. The star continued to survive, now with a black hole for a heart.

The black hole created in the centre of the star kept its angular momentum, causing matter to orbit it in smaller and smaller circles. The result was an accretion disk where gas orbits at nearly the speed of light. The heat generated by friction and collisions between particles in the accretion disk made it incredibly hot, heating it up to millions of degrees. This heat restricted how much the black hole could consume, just like the core of stars, the superhot material created radiation that blew away most of the food within its reach.

Black Hole Stars were only possible during a short window of time in the early universe, but if they existed, they would solve one of the largest mysteries of cosmology. These beasts of the early universe were excessive in every way, with a scale that was beyond words. Over 800,000 times wider than our sun, 380 times larger than the largest star we know today, and far below its surface was a black hole, growing rapidly as it consumed matter.

Black Hole Stars break everything we know about stars and their formation. They were only possible in a unique environment that will never exist again, and their existence would answer some of the most significant questions in cosmology. These cosmic wonders are truly a testament to the diversity and majesty of the universe, and their legacy continues to inspire and amaze us to this day.

"Surviving the Unsurvivable: Life After a Super volcano Eruption"

Earth is a hot ball of semi-molten rock with a heart of iron as hot as the surface of the sun. Heat from the planet's formation and the decay of radioactive elements pushes up to the surface through currents of rock. The Earth's crust is all that stands in the way of this heat, but it's only a fragile barrier like an apple skin around a blazing behemoth. And sometimes, this barrier can be broken, resulting in super volcanoes, the most catastrophic of all volcanic eruptions.

Volcanoes have two main sources - the boundaries between tectonic plates and mantle plumes. Tectonic plates are pieces of the crust that cover the Earth like a puzzle, moving against each other at up to 15 cm per year. When one plate wins, it becomes a new mountain range, and the loser plate is shoved underneath into the hot rock of the asthenosphere. This hot rock can trigger the melting of rock into magma, which rises to the surface and pierces through the crust, resulting in a volcano. On the other hand, mantle plumes are columns of hot rock that rise from the Earth's core to the surface, creating volcanoes in the middle of nowhere.

Volcanologists have created a logarithmic scale called the Volcanic Explosivity Index (VEI) to measure the volume of an eruption. VEI 2 eruptions are relatively small, while VEI 3 eruptions are devastating. VEI 5 eruptions are catastrophic, while VEI 6 eruptions can change the world. The most powerful of all are VEI 7 eruptions, known as super-colossal eruptions, which have only been encountered a handful of times in human history.

The term "super volcano" is not a scientific term, but it refers to the most massive of all volcanic eruptions. These eruptions can release thousands of times more energy than nuclear weapons and subject the climate to centuries' worth of change in a single year. The ash and gases from these eruptions can drown continents and create famines and centuries-long cold periods.

One such super volcano is the Yellowstone Caldera in Wyoming, USA. If this massive volcano were to blow, the ash and debris could reach as far as Europe, and the explosion would be equivalent to thousands of nuclear bombs. The ash and gas from this eruption could block the sun and lead to a volcanic winter, and the eruption could trigger earthquakes and tsunamis.

In conclusion, super volcanoes are a reminder of the massive power that lies within the Earth's crust. While we may never know when one of these behemoths will awaken, it's important to understand their potential impact and plan accordingly. The next super volcano eruption could bring an end to our current civilization, and it's up to us to be prepared.

"The Tug of War Between Security and Liberty"

errorism is a daunting reality that has forced people to demand more security measures in order to reduce their fear of being attacked. However, in the pursuit of safety, civil liberties have been significantly eroded, and government agencies have been given the power to conduct mass surveillance on citizens and collect their personal information. The aftermath of 9/11 saw the US government conclude that the law was not keeping pace with technology and created the Terrorist Surveillance Program to intercept communications linked to Al-Qaeda. However, these new powers were soon used to prove guilt by association and resulted in the FBI using immigration records to identify Arab and Muslim foreign nationals in the US, leading to the imprisonment of over 5,000 people without finding a single terrorist.

Edward Snowden's leak in 2013 revealed the extent of the NSA's data collection, with the agency demanding information about users from companies like Microsoft or Google, in addition to their daily collection of data from civilian internet traffic. The collection of massive amounts of data from everyone instead of focusing on criminals has not helped the NSA to stop any major terror attack. The NSA's success has come from classic target surveillance and not from their surveillance program. Spy agencies are also pushing to cripple encryption, which has resulted in a debate about privacy and security.

People are often told that if they have nothing to hide, they have nothing to fear, but this reasoning only creates a climate of oppression. Wanting to keep certain parts of one's life private doesn't mean that they are doing anything wrong. If a wrong person gets access to all of the data, the damage they could cause would be significant. Democracy requires democratic oversight, and anti-terrorism laws allow authorities to investigate and punish non-terrorism-related crimes more aggressively. If law enforcement is given powerful tools, they will use them, and this has been proven by the restriction of freedom of assembly and speech in Spain, Hungary, and Poland, and the restriction of freedom of expression and the press in Turkey.

The erosion of rights and mass surveillance have not led to any significant success in the fight against terrorism, but they have changed the nature of society. The problem of terrorism is complex, and no security apparatus can prevent people from building a bomb in their basement. It's important to keep the principle of proportionality in mind and understand that creating master keys to enter millions of phones is not the same as searching a single house. To make full use of existing potential, better international cooperation and more effective security and foreign policies are required, along with better application of present laws instead of new and stricter ones that undermine freedom.

In conclusion, the pursuit of safety should not be at the cost of destroying the fundamental rights and liberties that people hold dear. The fight against terrorism requires a proportionate response that doesn't undermine the principles of democracy and personal freedom. People need to be vigilant and protect their rights, ensuring that they are not taken away in the name of security.

"The Web Weaved: The Rise of the Internet and Its Inventor"

The Internet has revolutionized the world as we know it, connecting people from all corners of the globe, changing the way we communicate, do business, and access information. But who exactly invented the Internet? The answer to this question is not as simple as you may think. The Internet is not a single invention created by one person, but rather a combination of numerous technological advancements and innovations brought forth by scientists, researchers, and engineers from all over the world.

The Internet as we know it today has its roots in the 1960s with the ARPANET, a computer network experiment developed in the United States. The aim of the ARPANET was not to create a communication network but to optimize processor usage or time-sharing, which allowed scientists to share computer power. This was because before the 1960s, there were no computer networks, and big machines known as mainframes processed computing tasks one at a time. With time-sharing, these machines could process several tasks at once, and scientists could use their power. This sparked the idea of connecting computers together to make communications easier.

While the ARPANET was underway in the United States, similar efforts were underway in other countries. In Britain, for instance, the National Physical Laboratory developed a commercial network, but it failed to take off due to a lack of funding. However, they came up with the idea of packet switching, a method of avoiding congestion in busy networks by cutting up data at one end and putting it back together at the other. The French also played a role in the development of the Internet. They were working on a scientific network called

CYCLADES, but with limited resources, they focused on direct connections between computers, rather than working with gateway computers.

By the early 1970s, computer infrastructure had been developed, but communication was still awkward and patchy because different networks couldn't talk to each other. This problem was solved with the invention of the TCP/IP protocols, which form the basic communication language of the Internet, ensuring that data is properly labelled and reassembled even if some pieces take a different route. In 1975, networks began communicating with each other, and the Internet as we know it today was born. Another important development was email, which was developed for the ARPANET in 1972 and quickly became the most used form of communication on the Internet.

In the 1980s, Tim Berners-Lee, a British engineer, played a critical role in the development of the Internet by inventing an interface using HTTP, HTML, and URLs that made internet browsers possible. He created the World Wide Web, which made it easier for scientists to manage their information and share and interconnect their work. The first website was created at CERN in France in 1991, and the Internet continued to expand rapidly and steadily, becoming accessible to the masses around 1995.

It's important to note that the Internet exists because we need to communicate, and it's a natural evolutionary step in our quest to connect with one another. It wasn't invented by one person, but by a collective of brilliant minds from all over the world who put the building blocks in place to create the powerful tool we use today. While it's often said that US Vice President Al Gore "invented the Internet," this is not true. However, he did play a role in promoting legislation that encouraged the spread of the Internet.

In conclusion, the Internet is a product of a global effort by scientists, engineers, and researchers who worked together to make communication easier and more accessible. It's not the creation of one person, but the result of a combination of technological advancements and innovations from around the world.

"From Red to Green: A Guide to Terraforming Mars with Lasers"

The idea of terraforming Mars into a green and habitable world may no longer be just a far-fetched science fiction tale. With humanity's advancements in space technology and laser technology, it's becoming a possibility, albeit on a large timescale.

To terraform Mars, we first need to tackle its pressing problems: a dry and barren landscape with no soil, a thin atmosphere that is inhospitable to life, and high levels of radiation. Our goal is to create a new home for humanity by giving Mars a proper atmosphere similar to Earth's, composed of 21% oxygen, 79% nitrogen, and a tiny bit of CO_2, at an average temperature of 14°C and under 1 bar of pressure.

To achieve this, we must first melt Mars' surface and release the bound gases in its rocks, such as oxygen in iron oxides and carbon dioxide in carbonates. This can be done by using a powerful laser, such as a solar-pumped laser that runs continuously and is powered directly by sunlight. By melting the surface, about 750 kg of oxygen and carbon dioxide will emerge from every cubic meter of rock.

The resulting inferno would be a sight to behold, with skies shrouded in storms and the ground glowing red-hot from currents of lava, but it would also release all the water in the polar ice caps and underground, forming clouds that rain down over the entire planet. This rain would wash out the harmful gases from the atmosphere and carry away harmful elements that accumulated on the surface, eventually forming shallow oceans.

After 50 years of continuous lasering, we would have an atmosphere nearly 100% oxygen but lacking nitrogen. To make it similar to Earth and safer, we would have to import nitrogen from

Titan, a large moon of Saturn covered in a thick atmosphere that's almost entirely nitrogen.

With a proper atmosphere in place, the next step is to create oceans and rivers, weather the ground into fertile soil, and install a biosphere on the surface. This biosphere would need to be protected with measures that can stand the test of time, to prevent the atmosphere from being undone.

In conclusion, while terraforming Mars is a complex and challenging process, it's becoming increasingly possible with advancements in technology and laser technology. With a little bit of hard work, determination, and a lot of laser power, we could turn Mars into a green and thriving new world for humanity.

"The Dark Dominion of the Slave Ants"

Ants have conquered the planet for over 100 million years, thanks to their perfected chemical communication and social behaviour. However, among the millions of ant species, there are around 50 species that practice slavery - the most extreme division of labour. The most intense of these slaver ant tribes is Polyergus, known for their specialized enslavement practices.

Polyergus ants are 4 to 10 millimetres long, with brown to blackish bodies and sickle-shaped mandibles. They are known to only exist for raiding, as they have lost the ability to care for themselves. In a Polyergus colony, slaves make up 80 to 90% of the ants, with a few hundred Polyergus and a single queen controlling thousands of slave ants.

The story of Polyergus begins on a mild summer morning on a sunny field, where a colony of over ten thousand Formica ants build a thriving underground nest. A lone Polyergus scout briefly shows up, before returning with a raiding party of a thousand warrioresses. The attack begins with the clearing of the nest's entrance, followed by hundreds of attackers rushing inside. The Polyergus use their mandibles to pierce and kill, and also release a propaganda pheromone that makes the defenders confused and unable to mount a defence. The attackers then search for the colony's babies, taking hundreds of pupae and larvae back to the Polyergus colony.

Once the stolen Formica offspring arrive at the slaver colony, they are covered in Polyergus pheromones, chemically imprinting on the Polyergus as though they are part of the colony. The new slaves then begin to work for their masters unconditionally, keeping the nest clean, caring for the next generation of slaves and masters, hunting for food, and so on.

The story of the slaver ant is a horrific reminder of the cruel banality of nature. The Polyergus raids are efficient and effective, as the attackers turn their victims into slaves through chemical warfare and brainwashing. The victims are unable to resist or mount a defence and are forced to serve their new masters until they die, only to be replaced by new victims harvested in brutal raids.

The Polyergus story is not just a tale of slavery and oppression, but also of evolution and adaptation. The ancestors of Polyergus may have started as regular ants, but as they began to abduct other ants, they lost the ability to collaborate and communicate with each other. They have since evolved into the ruthless slaver ants we know today, relying on chemical warfare and brainwashing to control their victims.

The horror of the slaver ant is a reminder of the brutal and unforgiving world of nature, where survival of the fittest is the only law. But it also serves as a warning, showing what can happen when social animals lose their ability to collaborate and communicate with each other. In a sense, the story of the slaver ant is a cautionary tale of what can happen when we become too focused on our own interests and lose sight of our connection to each other and the world around us.

The Horror of the Slaver Ant

"The Power of Friendship: A Solution to the Pangs of Loneliness"

Friends are an essential aspect of life that bring happiness, meaning, purpose, and security to our lives. Unfortunately, loneliness and disconnectedness are prevalent issues, particularly among young people. The global pandemic has made things worse, as social distancing has stopped people from interacting and forming new relationships. Making friends is a simple process, but it's not always easy. The key to making friends is spending time with people in the real world. This creates an opportunity for overlapping social circles and shared experiences to form. Proximity is even more important than shared interests.

One of the main reasons people lack friends is that they don't prioritize their friendships enough. Work, family, and other life events take up most of their time and energy, leaving little room for socializing. Furthermore, the average American teenager spends more time on social media than socializing with friends. Established friendships need time and attention to maintain and neglecting them can lead to a slow decline in the relationship.

Another issue is the structure of friendship networks. The friendship paradox states that most people have fewer friends than their friends, meaning that our friend networks are often built around central hubs. This can lead to a distorted self-perception and a sense of loneliness when central people disappear from our lives. This phenomenon can become more pronounced with big life events like moving to a new place or changing jobs.

It's never too late to make new friends and find a sense of connection. Don't compare yourself to others and focus on finding what works best for you. Building and maintaining friendships takes effort, but the reward is worth it. The relationships we form with friends provide us with a sense of community, happiness, and purpose that is essential to our well-being.

In conclusion, making friends is a simple process that requires regular time and effort. The key to building relationships is spending time with people in the real world and making an effort to stay connected. Don't let the structure of friendship networks or life events get in the way of finding happiness through friendships. Embrace the process and enjoy the journey of making friends.

"Deja Vu All Over Again: ...And We'll Do It Again"

Science communication is a crucial aspect of educating the masses about the advancements and knowledge gained through scientific pursuits. However, this process often requires the use of "lies to children", meaning a useful oversimplification of complex subjects. This is done to make it easier to grasp the concept and build a foundation for further understanding. The idea behind this process is that as the understanding grows, so does the complexity of the subject, thus eventually leading to a deeper understanding of reality.

The fast-moving world of the information age makes it difficult to have an expert-level understanding of every field of study. Hence, the need for summaries and explanations that provide a solid overview without overwhelming the audience with too much detail. The challenge lies in finding the right metaphors and stories that can help convey the true nature of things while using a language that our brains can understand.

Physics, for example, often uses colourful depictions of quarks to explain their different spins. While these simplifications help to visualize the relationship between quarks, they can also create a false image in our minds of reality. Similarly, molecules are depicted as neat diagrams in textbooks, but in reality, they are nothing like what we imagine them to be. Instead, they are buzzing, vibrating entities held together by something called charge, a concept that is yet to be fully understood.

Simplifications like these are not just meant for the layman but also for experts. Scientists often use simplified models as a basis for discussions with colleagues from different fields. However, this oversimplification of science can also have negative consequences. The true nature of reality is complex, and the universe doesn't care if we understand it. Science is a process that works towards gaining knowledge, not an absolute truth generator. Hence, when oversimplifications are presented as absolute truths, it can lead to a false sense of security and an illusion of deep knowledge. This can lead to bad decisions, especially when people ignore actual experts and rely on their gut feelings.

In conclusion, the process of science communication is a delicate balance between oversimplifying and retaining the nuance and complexity of the subject. It is essential to find the right metaphors and stories that can help convey the true nature of things and prepare the ground for further explanations. Science communication is not just about enlightening the ignorant but necessary for the progress of our species as a whole. The simplification of science can be problematic, but if done correctly, it can help us better understand our world and make more informed decisions based on facts and testable ideas.

Did you love *The Paradox Explorer*? Then you should read *Life Illumination: 100 TEDx Talks to treasure*[1] by Rohan Aggarwal!

[2]

"Unlock the power of innovative ideas and groundbreaking insights with this comprehensive collection of 100 TEDx Talks. This book brings together the most inspiring voices of our time to share their stories, passions, and vision for a better world. Whether you're seeking inspiration for your own life or looking to spark a revolution, this book is the ultimate source of wisdom, creativity, and inspiration.

"Dive into a diverse array of topics, from science and technology to art and social change, and be captivated by the unique and powerful perspectives of the world's most innovative thinkers. Each talk has been carefully curated for its impact and relevance, providing a

---

1. https://books2read.com/u/bwyyaG

2. https://books2read.com/u/bwyyaG

thought-provoking and empowering perspective on the world we live in.

With "Life Illumination: 100 TEDx Talks to treasure", you'll gain access to a wealth of knowledge and unlock the potential for positive change. Whether you're looking to improve your relationships, achieve your goals, or simply find more joy and fulfillment in life, this book offers practical, actionable advice for developing a positive mindset and overcoming the obstacles that hold you back.

"So, sit back, get ready to be inspired, and let the ideas of the world's most visionary thinkers change your life forever. Order your copy of "Life Illumination: 100 TEDx Talks to treasure" today and discover the power of great ideas!"

Also by Rohan Aggarwal

The Paradox Explorer